Christoph Baumgartner

Clinical Electrophysiology of the Somatosensory Cortex

A Combined Study Using Electrocorticography, Scalp-EEG, and Magnetoencephalography

Springer-Verlag Wien New York

Dipl.-Ing. Dr. med. Christoph Baumgartner
Neurological University Clinic
Vienna, Austria

Sponsored by "Fonds zur Förderung der Wissenschaftlichen Forschung"

Printed in Austria
Printed on acid-free paper

Typeset by Bernhard Computertext KG, A–1030 Wien
Printed by Eugen Ketterl GmbH, A–1180 Wien

With 53 Figures

ISBN 3-211-82391-3 Springer-Verlag Wien New York
ISBN 0-387-82391-3 Springer-Verlag New York Wien

Preface

The functional anatomy of human somatosensory cortex is of both scientific and clinical interest. Scientifically, it provides insights in information processing in the human brain. Clinically, it helps to avoid neurological deficits by sparing essential brain regions during neurosurgical procedures adjacent to central fissure.

In the present study the functional organization of the human somatosensory cortex was investigated with electrophysiological techniques using a combined approach of cortical stimulations and somatosensory evoked responses on electrocorticography, scalp-EEG, and magnetoencephalography. The spatiotemporal structure of the evoked response was studied with biophysical modeling techniques which allowed identification of the three-dimensional intracerebral location, time activity, and interaction of the neuronal sources activated following peripheral somatosensory stimulation. Furthermore, the somatotopic organization of hand and lip somatosensory cortex was investigated. The relative value of invasive (cortical stimulations and electrocorticography) and non-invasive (scalp-EEG and magnetoencephalography) was assessed. The combined use of scalp-EEG and magnetoencephalography was useful to increase non-invasive localization accuracy.

I want to thank several people who significantly contributed in completion of the present work. *Univ.-Prof. Dr. Lüder Deecke,* Chairman of the Neurological University Clinic, Vienna, Austria, supported me throughout my career at the Neurological University Clinic in Vienna since 1985. *Dr. William W. Sutherling,* Associate Professor at the University of California, Los Angeles, who was my advisor during my stay from 1987–1989 at the Department of Neurology, University of California, Los Angeles where most of the present work was done. He has taught me clinical neurophysiology including magnetoencephalography and cortical stimulations on epilepsy patients with chronically indwelling subdural grid electrodes and rouse my interest in the somatosensory system. I want to mention that the section on somatotopy of human hand somatosensory cortex as studied on ECoG represents primarily his research effort (Sutherling, W. W., Levesque, M. F., and Baumgartner, C. Cortical sensory representation of the human hand: Size of finger regions and nonoverlapping digit somatotopy. *Neurology* 42: 1020–1028, 1992). He kindly let me use these results in the present study. *Dr. Daniel S. Barth,* presently Assistant Professor at the University of Boulder, Colorado, who has shaped my attitude towards neuroscience. I want to thank him for numerous stimulating discussions where he short-circuited my brain. *Herr Raimund Petri-Wieder* and *Herr Edwin W. Schwarz* supported the publication of this study by *Springer-Verlag Wien – New York.*

Finally, I want to thank the *Fonds zur Förderung der wissenschaftlichen Forschung of Austria* for financial support for a two years stay at the Department of Neurology, University of California, Los Angeles with an Erwin Schrödinger Auslandsstipendium (J246M and J334MED). I could not have continued my work after my return to Austria without the funds for a research project by the Fonds zur Förderung der wissenschaftlichen Forschung (Forschungsprojekt P7434). The printing of this study was made possible by financial support of the Fonds zur Förderung der wissenschaftlichen Forschung (D2120MED).

Christoph Baumgartner

Contents

1. Introduction

The cutaneous body surface is mapped in a one-to-one projection onto primary somatosensory cortex known as 'homunculus' representation [278, 279, 280]. Knowledge about the functional organization of primary somatosensory cortex is both of neurophysiological interest to understand sensory information processing in the human brain and of clinical interest to avoid neurologic deficits during neurosurgical procedures adjacent to central sulcus. Although animal studies [16, 17, 18, 182, 183, 184, 185, 186, 187, 188, 192, 193, 194, 244, 245] have provided important information concerning organization of primary motor and somatosensory cortex, these data cannot be readily extrapolated to the human brain as there are significant species-specific differences in the functional organization of these cortical areas [7, 145, 194]. Therefore, studies of the human somatosensory system itself are warranted.

This study deals with 2 fundamental questions concerning functional organization of human primary somatosensory cortex:

1. Functional anatomy of human somatosensory cortex.

We studied functional anatomy of human somatosensory cortex with cortical stimulations and somatosensory evoked responses recorded on electrocorticography (ECoG), on scalp-EEG, and on magnetoencephalography (MEG).

2. Somatotopic organization of human somatosensory cortex.

We studied somatotopic organization of human somatosensory cortex with cortical stimulations and somatosensory evoked responses recorded on electrocorticography (ECoG), on scalp-EEG, and on magnetoencephalography (MEG).

1.1. Functional Anatomy of Human Somatosensory Cortex

1.1.1. Clinical Importance

An understanding of functional anatomy of human somatosensory cortex requires knowledge about its exact anatomical location as well as about spatiotemporal features of sensory information processing. Exact localization of primary somatosensory cortex is of direct clinical importance. Patients undergoing neurosurgical procedures adjacent to central fissure for various reasons (e.g. epilepsy surgery, resection of tumors or arteriovenous malformations etc.) are at risk to develop disabling and irreversible neurologic deficits (e.g. severe paresis) if essential cortex is resected [5, 218, 228, 378]. On the other hand, the main goal of epilepsy surgery is complete removal of the epileptogenic tissue which sometimes is close to central fissure [119, 131, 378]. Furthermore, incomplete removal of tumors may result in relapses and therefore unsatisfactory results [378]. However, anatomical criteria are insufficient to identify central fissure in up to 50% of the cases [378]. This is especially true during surgical procedures where the brain is exposed only to a very limited degree and is covered by the pial membrane, which makes identification of fissures and gyri significantly more difficult [226]. Furthermore, lesions may displace common anatomical landmarks and cause functional reorganization of the cortex [149]. Exact functional localization of primary somatosensory and somatomotor cortex in these patients therefore could lead to better surgical outcomes concerning the underlying disease and could help to avoid neurological deficits induced by the operation.

1.1.2. Methods to Study Functional Anatomy of Human Somatosensory Cortex

In humans, functional organization of primary somatosensory cortex can be studied by direct electrical stimulation of the cortical surface and by somatosensory evoked responses (SERs) which can be recorded from the surface of the brain on electrocorticography (ECoG) or non-invasively from the scalp on electroencephalography (EEG) and on magnetoencephalography (MEG).

1.1.2.1. Direct Cortical Stimulations

Direct electrical stimulations of the cortical surface are presently considered the gold standard in cortical localization [152, 218, 219, 228, 229, 261, 262]. Cortical stimulations can be performed either intraoperatively or from chronically indwelling subdural grid electrodes. Intraoperative stimulations necessitate the use of local anesthesia as general anesthesia precludes the ability of the patient to report sensory experiences induced by the stimulations [5, 378]. Additionally, general anesthesia, due to the use of muscle relaxants, usually makes it difficult to elicit discrete movements during stimulation of primary motor cortex [5, 146, 149]. However, performing brain surgery under local anesthesia imposes stress both on the patient and the surgical team [5]. Especially, in patients with mass lesions, opening of the cranium in an awake patient may be hazardous due to increased intracranial pressure [378]. Furthermore, cortical stimulations performed intraoperatively suffer from time limits and can prolong operation time significantly [378].

Stimulations from chronically indwelling subdural grid electrodes, on the other hand, are not restricted by time limits and therefore allow more detailed and sophisticated protocols for cortical stimulations [218, 219, 228,

229]. However, placement of these grids requires an additional craniotomy and therefore is only indicated in patients who are candidates for epilepsy surgery with an epileptogenic focus adjacent to central sulcus or to differentiate a temporal and frontal seizure focus [228, 345, 347]. In these patients, chronically indwelling grids are also used to record interictal as well as ictal epileptiform activity and thus allow exact delineation of the seizure focus.

Both intraoperative and extraoperative cortical stimulations helped to get a better understanding of functional brain topography in humans and allow highly selective stimulations of circumscribed brain regions [219, 228, 229, 261, 262, 278, 279, 280]. However, cortical stimulations also have several disadvantages specifically for the investigation of the somatosensory cortex. First, they rely on the patient's reports of sensory experiences and therefore are subjective. Second, they may yield ambiguous results as stimulation of primary motor cortex sometimes results in sensory phenomena, and stimulation of primary sensory cortex sometimes results in motor phenomena. Third, they usually result in sensory phenomena in widespread areas of the body surface [146, 149, 378]. Finally, motor cortex may prove to be electrically unexcitable in small children [146, 226].

In the present study, we investigated 6 epilepsy patients who were evaluated with chronically indwelling subdural grid electrodes for a definitive localization of the seizure focus. We used cortical stimulations to map out essential cortical areas, i.e. primary motor and somatosensory cortex, and language-related cortex. We compared the results from cortical stimulations with source localizations obtained from somatosensory evoked potentials recorded from these grids.

1.1.2.2. Somatosensory Evoked Potentials on Electrocorticography

An alternative method to study human somatosensory cortex are somatosensory evoked potentials (SEPs) during peripheral nerve stimulation recorded directly from the surface of the brain on electrocorticography (ECoG) [5, 8, 11, 32, 146, 149, 226, 231, 346, 349, 378]. SEPs recorded on ECoG have several advantages compared to direct cortical stimulations. First, SEPs yield – besides spatial information like cortical stimulations – also temporal information and therefore allow investigation of spatiotemporal organization of human somatosensory cortex. Second, SEPs are objective as they do not rely on the patient's subjective reports. Finally, SEPs can be recorded under general anesthesia because short latency SEPs are not significantly influenced by various types of anesthesia [5, 378].

SEPs recorded on ECoG are considered most accurate for localization of sensorimotor cortex because ECoG neither is distorted by the resistive properties of the skull like scalp-EEG nor loses spatial resolution due to distance like scalp-EEG and MEG [1, 72, 96, 287, 346]. SEPs on ECoG can be performed either intraoperatively or extraoperatively form chronically indwelling subdural grid electrodes. Intraoperative SEPs on ECoG are accurate in localizing hand sensory cortex [378], but can increase intraoperative time by 15 to 45 minutes (due to requirement of multiple placements of the electrode grid to screen for the maximum electrical activity) and in some cases cannot identify central fissure (due to its location outside the operative field). Therefore, intraoperative SEPs on ECoG are sometimes difficult to perform precisely in the operating room, especially when recording of epileptiform activity has to be achieved during the same surgical procedure [231]. Extraoperative SEPs on ECoG from chronically indwelling subdural grid electrodes are not restricted by time constraints, and therefore can be performed more precisely. Placement of these grids, however, necessitates an additional craniotomy. As already mentioned (Section 1.1.2.1), the indication for this technique therefore is restricted to epilepsy surgery candidates where the exact localization of epileptogenic tissue and of essential cortex cannot be achieved with other less invasive techniques.

Woolsey and colleagues pioneered the study of evoked potentials recorded from the surface of the brain in animals and in man [380, 381, 382, 383, 384]. Jasper et al. [181] compared cortical stimulations and SEPs and concluded that SEPs might be a useful alternative method of localization when subjective responses to electrical stimulation are ambiguous. Since these pioneering studies, several investigators have recorded evoked potentials from the surface of the brain [4, 5, 7, 8, 11, 32, 49, 50, 51, 59, 145, 146, 149, 172, 204, 231, 248, 275, 335, 346, 349, 374, 378]. However, definitive guidelines for localization of hand primary motor and somatosensory cortex have been provided only by few studies [5, 32, 146, 149, 378]. In some patients with large lesions, however, SEPs are non-localizing or unobtainable by conventional procedures. Therefore, more sensitive and accurate localization procedures would be useful.

In the present study, we recorded SEPs on ECoG from chronically indwelling subdural grid electrodes in 6 epilepsy patients who were evaluated for a definitive localization of the seizure focus. We applied multiple dipole modeling to clarify the neuronal sources underlying SEPs and thus tried to provide more sensitive and objective criteria for localization of central fissure. Finally, we compared the results with those from cortical stimulations and non-invasive recordings on scalp-EEG and MEG.

1.1.2.3. Somatosensory Evoked Potentials on Scalp-EEG

Cortical stimulations and somatosensory evoked potentials (SEPs) recorded on ECoG are invasive procedures and therefore are inappropriate for routine patient evaluation. Scalp-EEG, on the contrary, can be performed non-invasively and therefore can be used to investigate non-surgical patients and to study functional anatomy of somatosensory cortex in normal controls. The disadvantage of scalp-EEG is that electrical potentials are attenuated and smeared by the resistive properties of the skull and scalp, and that measurements are made at a greater distance from the source resulting in a poorer spatial resolution [1, 72, 96, 287].

Scalp-recorded human SEPs were introduced by the pioneering contributions of Dawson using superimposition [85, 86] and, later, an electromechanical summating ('averaging') technique [87]. Since then, scalp-recorded SEPs have been widely used both in the evaluation of patients to assess various abnormalities of the sensory pathway as well as in normal subjects to study the physiology of the somatosensory cortex [65, 98, 125, 130, 332, 388, 389, 390].

In many initial studies as well as in clinical routine, SEPs were recorded only from a few number of electrodes and the main interest was focused on peak latencies or amplitude criteria. However, peripheral nerve stimulation usually elicits a complex sequence of waveforms which can be recorded over the entire scalp [106, 142, 143]. Thus, the interpretation of single SEP traces may be biased by the electrode sites or montages selected [100]. Furthermore, no single electrode montage can disclose all pertinent SEP features [106, 109, 143]. Therefore, localization of somatosensory cortex and study of its functional anatomy requires simultaneous recording from a large number of electrodes. Multichannel recordings of SEPs were pioneered by several investigators [7, 73, 74, 101, 106, 120, 139, 142, 143]. These recording strategies thus yield comprehensive information, but on the other hand confront the investigator with a large amount of data. Thus, brain mapping techniques which have been used in various fields of EEG research [54, 122, 214, 296] also were applied to the investigation of scalp SEPs [100, 108]. These techniques allow a graphical display of the data and have been used to re-assess the spatiotemporal information of SEP components [94, 95, 100, 108, 109, 110, 354].

Furthermore, numerous SEP studies have raised the question on the neuronal generators of different SEP components [4, 7, 100, 106, 239]. Knowledge of neuronal sources underlying SEPs would enhance the diagnostic significance of SEPs in patient evaluation and would especially be useful to get non-invasive localization estimates of somatosensory cortex. So far, most investigators studying the neuronal sources of scalp SEPs based their conclusions on analysis of waveforms, inspection of brain map features, or lesion studies [94, 95, 98, 100, 109, 238, 239, 354]. However, these techniques do not take into account volume conduction and the biophysical laws of EEG generation. Thus – despite a large number of studies in this field – the exact anatomical location of the neuronal generators underlying scalp SEPs is still controversial. Furthermore, no objective criteria for localization of somatosensory cortex as derived from scalp-EEG have been provided so far. Source localization techniques based on biophysical models should be useful to obtain additional information on the neuronal sources of SEPs and to provide accurate localization estimates of central sulcus.

Therefore, we recorded SEPs on scalp-EEG in 4 normal subjects and one epilepsy patient. We applied multiple dipole models to study the three-dimensional locations and time activities of the neuronal sources underlying SEPs, and thus tried to achieve criteria for non-invasive localization of human hand somatosensory cortex. We compared these results with invasive recordings on ECoG and with non-invasive recordings on MEG.

1.1.2.4. Somatosensory Evoked Fields on Magnetoencephalography

Magnetoencephalography (MEG) is a new neurophysiological technique which allows the measurement of the weak magnetic fields induced by neuronal currents. In contrast to electric fields, these magnetic fields are virtually not distorted by the skull and scalp which results in a better spatial resolution of MEG compared to scalp-EEG [81, 162, 198]. Therefore, somatosensory evoked magnetic fields (SEFs) recorded on MEG should be useful to study the neuronal sources of the human evoked response and to get accurate non-invasive localization estimates of somatosensory cortex. SEFs have been studied using both the steady state paradigm and transient evoked responses. The basic concept underlying the steady state paradigm is that each neural network can be considered to possess a specific resonance frequency which is determined by its intrinsic organization [30, 133]. Driving the system at its resonance frequency – which lies in the alpha-theta range for the somatosensory system – results in so-called steady state evoked responses. Transient evoked responses, on the contrary, are based on the fact that peripheral sensory stimulation results in synchronized cortical activity which can be segregated from the ongoing brain activity by averaging and thus gives rise to transient somatosensory evoked magnetic fields which can be measured on MEG [162, 258].

The steady state paradigm allowed rapid accumulation of high-quality data even in a noisy environment. Thus, Brenner et al. [48] were the first to report on steady-state MEG responses during median nerve stimulation indicating a current source in contralateral primary somatosensory cortex. Similarly, Kaufman et al. [197] and Okada et al. [264] could reproduce these findings using transient evoked responses and suggested an origin of the somatosensory evoked magnetic fields at latencies of 20-250 ms in the vicinity of the central sulcus. Since then somatosensory evoked magnetic fields have been recorded in several studies using both the steady state paradigm [254, 268, 342] or transient evoked responses [34, 42, 166, 176, 200, 222, 309, 311, 346, 352, 353, 375]. In contrast to steady state responses, transient evoked responses also allow investigation of the temporal evolution of information processing in somatosensory cortex and thus are more appropriate to study its functional anatomy.

Many previous MEG studies have used dipole modeling to study the neuronal sources underlying SEFs [166, 176, 200, 222, 254, 268, 309, 311, 342, 346, 352, 353, 375]. Especially, the studies which compared MEG results to those obtained from electric recordings could resolve some ambiguities concerning the neurogenesis of evoked responses [346, 375]. This could be expected from theoretical considerations as electric and magnetic recordings yield both confirmatory and complementary information to each other [69, 81]. Furthermore, MEG localization estimates for central fissure yielded promising results in comparison with invasive recordings.

In the present study, we recorded SEFs in 9 normal subjects and used multiple dipole modeling to study the spatiotemporal structure of neuromagnetic evoked responses. We compared the results with those obtained from electric recordings on ECoG and scalp-EEG, and thus tried to improve criteria for non-invasive localization of somatosensory cortex.

1.1.3. The Neuronal Sources of Somatosensory Evoked Responses

As already mentioned, an ultimate understanding of somatosensory evoked responses (SERs) requires quantification of the dynamic physical activity of their underlying neuronal sources [7]. Despite a large number studies, there still exist controversies concerning the number, exact anatomical location, and time activity of the brain regions activated during peripheral nerve stimulation [4, 7, 8, 32, 34, 42, 94, 100, 105, 109, 231, 239, 346, 352, 375]. Specifically, it is not clear whether the neuronal generators of SERs are located in postcentral gyrus and thus in primary somatosensory cortex exclusively or if there exist additional precentral generators in somatomotor cortex contributing significantly to the evoked responses. Thus, some authors believe that there are two sources in postcentral gyrus, i.e. one tangential source in the posterior bank of central sulcus (area 3b) and one radial source in the anterior crown of postcentral gyrus (area 1) [4, 7, 8, 32, 34, 42, 49, 50, 51, 346, 375, 378]. On the contrary, other investigators proposed an additional radial source in precentral gyrus (area 4) [59, 100, 105, 109, 145, 231, 239, 335].

Dipole models allow to infer objectively the three-dimensional location and time activity of the neuronal sources generating SERs as they are based on biophysical laws. Although dipole models have been introduced in EEG research more than 10 years ago [83, 84, 202, 372], their application to SERs on ECoG and scalp-EEG has been rather scarce [7, 32, 346]. On the contrary, the dipole approach has been widely used for studying SERs on MEG [166, 176, 200, 222, 254, 268, 309, 311, 342, 346, 352, 353, 375] which also should inspire EEG research.

However, these studies used single equivalent dipole models and restricted their analysis to selected time points or intervals given by peaks or constancy of field distribution. Thus, not the full spatiotemporal information provided by SERs was taken into account. Evidence from animal and human studies suggests that human SERs are generated by multiple simultaneously active brain regions producing electric and magnetic fields that overlap both in space and time [7, 8, 100, 193, 346, 352, 375]. Modeling of activity generated by multiple brain regions with a single dipole can lead to false localization estimates [40, 259]. Indeed, single dipole models often yield inadequate fits for modeling the human somatosensory evoked response [346].

In the present study, we therefore applied novel source localization techniques which allow modeling of multiple simultaneously active brain sources over the entire time domain. Thus, we could use the full spatiotemporal information contained in the SERs. We applied this approach to SERs recorded on ECoG, scalp-EEG, and MEG and thus tried to study the number of underlying neuronal sources, their exact anatomical location, their time course, and the spatiotemporal structure of their fields. Details of this modeling procedure are presented in Section 3.

1.2. Somatotopy of Human Somatosensory Cortex

1.2.1. Current Concepts Concerning the Somatotopic Organization of Somatosensory Cortex

Many of our current concepts concerning the organization of human primary somatosensory cortex stem from the landmark studies on electrical stimulations of the cortical surface performed by Penfield and colleagues [278, 279, 280] and on evoked responses recorded from the surface of brain performed by Woolsey and coworkers [380, 381, 382, 383, 384].

Penfield and colleagues reported a somatotopic representation of the body surface in human primary somatosensory cortex based on direct cortical stimulations performed in the operating room [278, 279, 280]. They described a detailed somatic sequence of sensory representation in postcentral cortex: 'As far as sensation is concerned, the representation passes from trunk to neck to scalp and then, after the interruption of the upper extremity, it passes onto forehead, face, lips, and jaw. This is a continuous progression over top of head, and without reversal. From the lips, the representation enters the inside of the mouth and throat' [279]. Specifically, these authors reported a sensory sequence for the hand with the thumb most lateral inferior and a progression to medial superior in the order index finger, middle finger, ring finger, and little finger [278]. There were large representations of hand, face, and tongue, and small representations for leg, arm, and trunk resulting in a distorted cortical representation of the cutaneous body surface known widely as 'homunculus'.

Woolsey and colleagues based their studies on evoked potentials recorded from the cortical surface both in animals and man [380, 381, 382, 383, 384]. In these studies, cutaneous receptive fields related to specific cortical sites were defined by mechanical stimulation and thus figurine style results could be derived. These investigators concluded that there was a large single representation of the contralateral body surface within the anterior parietal cortex in macaque monkeys and man. Thus, the sensory cortical representation comprised Brodman's areas 3a, 3b, 1, and 2, and consisted of a homuncular representation with the foot located most medial superior, the face most lateral inferior, and the hand in between. Although a single homuncular body representation is still considered a valid working hypothesis in clinical neurology and neurosurgery, animal experiments have further refined our knowledge about organization of somatosensory cortex.

The important single unit experiments in cats and monkeys performed by Mountcastle and Powell evidenced a more complicated somatic sensory cortical representation providing evidence for a possible meaning of the cytoarchitectonical differentiation of somatosensory cortex [250, 251, 252, 253, 293, 294]. These authors found that cells in Brodman's area 3b (covering the posterior wall of the central sulcus) received input primarily from cutaneous receptors. Cells in area 1 corresponding to the anterior crown of postcentral gyrus were activated by both cutaneous and deep receptors. Finally, area 2 in the posterior crown of postcentral gyrus received input predominantly from deep receptors. Mountcastle and Powell noted that these 'submodal' inputs were segregated in a columnar fashion, i.e. in any given penetration perpendicular to the surface of the cortex, one response mode or the other was predominant. These findings were incompatible with a single homuncular cortical representation. These investigators therefore introduced the concept of rostrocaudal bands extending across areas 3a, 3b, 1 and 2. All body surface locations would thereby be subserved by peripheral receptors of all classes, and areas 3a, 3b, 1, and 2 would be considered part of a single representation.

Paul, Merzenich, and Goodman were the first to demonstrate two complete hand representations in primary somatosensory cortex of Macaca mulatta using a microelectrode multiunit recording technique: one representation was within area 3b, and the other in area 1 [277]. The concept of multiple independent somatosensory representations was subsequently further refined by the studies of Kaas, Merzenich and colleagues who demonstrated four functionally distinct strips in somatosensory cortex of Macaca mulatta, two representing the cutaneous body surface in areas 3b and 1, and two representing deep body tissue in areas 2 and 3a [192, 193, 194, 244, 245, 256]. Furthermore Jones et al. [188] documented separate and topographically organized projections from the ventral posterior lateral nucleus of the thalamus to areas 3b and 1 of somatosensory cortex in primates.

However, there are species specific differences in the organization of somatosensory cortex and these results cannot be readily transposed to the human brain [194, 245]. Thus, further studies on the topographical dynamical organization of human somatosensory cortex are warranted which was the aim of the present study.

1.2.2. Clinical Importance

In the present study on somatotopy of human somatosensory cortex, we were especially interested in cortical hand and digit representation for several reasons. First, the hand is our most important body part to achieve somatosensory information. This is reflected by its large cortical representation. Especially, the thumb is represented in a large cortical area [278, 383]. Second, resection of hand somatosensory cortex can result in difficulties with fine finger movements [149]. Thus, hand primary somatosensory cortex is functionally important and should be spared during neurosurgical procedures adjacent to central fissure. Therefore, exact delineation of hand somatosensory cortex, i.e. representation of all the individual digits, is of direct clinical relevance. We also studied somatosensory lip cortex. Despite the large cortical face representation, systematic SEP studies on this question have been rather scarce [33, 146, 226]. Due to its bilateral cortical representation, somatosensory face cortex is functionally less important than somatosensory hand cortex [149]. However, exact localization of face somatosensory cortex could be helpful during neurosurgical procedures to get a lower border of hand representation. This might be especially of clinical relevance in patients with frontal lobe seizures arising from operculum.

1.2.3. Methods to Study Somatotopy of Human Somatosensory Cortex

In humans, somatotopy of somatosensory cortex can be studied by cortical stimulations and by somatosensory evoked responses recorded on ECoG, scalp-EEG, and MEG.

1.2.3.1. Direct Cortical Stimulations

As already pointed out in Section 1.1.2.1, cortical stimulations are considered the most accurate method for localizing essential cortical functions, have been widely used in neurosurgical patients to delineate focal excisional surgery, and have generated detailed somatotopic maps of somatosensory cortex [228, 261, 262, 278, 279, 280]. Some general disadvantages of cortical stimulations have been mentioned already: use of local anesthesia and time limits when performed acutely in the operating room; requirement of an additional craniotomy when performed from chronically indwelling subdural grids; subjectiveness as they rely on the patient's reports; occasionally ambiguous results when stimulation of postcentral gyrus elicits motor responses and stimulation of precentral gyrus elicits sensory responses. Besides these general disadvantages which equally apply to the localization of central fissure (Section 1.1.2.1), cortical stimulations also have special shortcomings for the study of somatotopy of human somatosensory cortex. First, cortical stimulations usually are non-physiological and may produce sensations in a wide area of the body surface. Especially the presently used macroelectrodes result in activation of an extended cortical area and thus do not allow a refined topographic analysis. Second, cortical stimulations do not provide information about the temporal evolution of information processing in somatosensory cortex. Third, cortical stimulations are highly non-physiological as they result in an uncoordinated activation of cortical cells both in the vertical and horizontal direction which is not comparable with the differentiated spatiotemporal activation of somatosensory cortex in response to physiological stimuli [111, 228]. Thus, the results of cortical stimulations can provide only a first, crude insight in somatotopic organization of human somatosensory cortex.

In the present study, we performed cortical stimulations from chronically indwelling subdural electrode arrays in 4 epilepsy patients undergoing presurgical evaluation. We localized primary motor and somatosensory cortex as well as language-related cortex. During stimulation of primary somatosensory cortex we paid specific attention to the areas involved in sensory experiences as reported by the patients and thus tried to map out a somatosensory homunculus. We compared the results of cortical stimulations with the source localizations obtained from SEPs recorded on ECoG in these patients.

1.2.3.2. Somatosensory Evoked Potentials on Electrocorticography

Somatosensory evoked potentials (SEPs) recorded on electro-corticography (ECoG) are an alternative method to study somatotopic organization of somatosensory cortex. The general advantages of SEPs have already been described in Section 1.1.2.2, i.e they provide spatial and temporal information about the somatosensory cortex, are objective, can be used under general anesthesia and have a high spatial resolution compared to scalp-EEG. Furthermore, SEPs on ECoG provide some special advantages for the study of somatotopy. First, different parts of the cutaneous body surface can be stimulated selectively which allows a refined investigation of somatotopic features. Second, while SEPs evoked by electrical peripheral nerve stimulation still represent non-physiological stimuli, they simulate physiological reality much closer than cortical stimulations and result in a well organized spatiotemporal activation pattern in somatosensory cortex.

As already mentioned above (Section 1.2.1), Woolsey and colleagues demonstrated in their landmark studies a topographic relationship between the cutaneous body surface and its cortical representation in animals and man [380, 381, 382, 383, 384]. However, the human studies were performed intraoperatively suffering from time constraints and therefore it was not possible to map more than a limited number of cortical sites in a given patient. Even in the patients studied most extensively, exploration of the body surface was incomplete [383]. Concerning cortical digit representation, these studies demonstrated a larger representation for the first three digits than for the last two digits. Somatotopy of digits was demonstrated across patient populations, but not separation of all 5 digits in the same human brain. Maximum SEP amplitudes were the criteria used to study somatotopic organization.

Since these classical studies, SEPs on ECoG have been used mainly to localize central sulcus by stimulating the median nerve [7, 8, 32, 51, 59, 145, 146, 149, 181, 204, 231, 335, 346, 378], but only few systematic studies on somatotopy have been performed [33, 348]. Despite the fact that the face, specifically the lips, is represented in a wide area of primary somatosensory cortex, reports on SEPs during face stimulation have been scarce. Lüders et al. [226] reported on SEPs recorded on ECoG during lip stimulation, but did not systematically investigate these data concerning their neuronal sources or concerning their topographic relationship to SEPs evoked during stimulation of other peripheral fields.

The aim of the present study therefore was twofold. First, we wanted to study somatotopy and spatiotemporal structure of human hand somatosensory cortex. Specifically, we investigated functional topography of the individual digits and the spatial relationship of cortical digit to median and ulnar nerve representations. We therefore recorded SEPs during stimulation of the median nerve, the ulnar nerve, and the individual digits in 4 epilepsy patients evaluated for a definite localization of the seizure focus using chronic subdural electrode arrays.

The second part of the study was devoted to the study of functional anatomy of lip somatosensory cortex. We inferred the question whether lip somatosensory cortex exhibited a similar spatiotemporal organization as hand somatosensory cortex. Furthermore, we investigated the topographic relationship of lip to hand somatosensory cortex. We therefore recorded SEPs from chronically indwelling subdural grid electrodes during stimulation of median nerve, ulnar nerve, and the lower lips in 3 epilepsy patients.

In both studies, we used dipole models to study the three-dimensional anatomical locations and time activities of the neuronal sources underlying SEPs during stimulation of different peripheral receptive fields. We validated our findings by cortical stimulations, by intraoperative photographs, and by neurological examinations after focal excisional surgery.

1.2.3.3. Somatosensory Evoked Potentials on Scalp-EEG

Scalp-EEG can be used easily to study physiological questions such as somatotopy of human hand somatosensory cortex in normal controls. Nevertheless, scalp-EEG studies reporting on this question have been rather scarce. Duff [121] studied SEPs during digit stimulation in normal controls using 36 closely spaced electrodes. He found that the location of peak P24 amplitude depended on the digit stimulated. With thumb stimulation peak P24 amplitude was located approximately two-thirds the distance from midline to central fissure, while stimulation of the other digits resulted in more medially located topographic peaks. He concluded that the resolution of topographic shifts by this technique was clearly limited as the topography of potentials evoked by index, middle, or little finger stimulation were frequently not distinguishable among intersubject recordings. In a recent study, Deiber et al. [94] recorded digit SEPs on scalp-EEG. These authors could not find a N20 scalp distribution that paralleled the somatotopic digit representation in primary somatosensory cortex. However, the putative dipole orientations were found to change according to the finger stimulated with the dipoles during thumb stimulation oriented most horizontally, and the dipoles during stimulation of the little finger oriented most vertically. On the contrary, for the P22 component a somatotopic arrangement of potential maxima was found. However, the results presented were pooled from 12 subjects, and digit somatotopy in a single brain was not shown. These studies based their conclusions on peak amplitudes or on inspection of bit-mapped color images, but did not use source localization techniques which take into account the laws of volume conduction and thus should be more accurate to study small somatotopic shifts of activated cortex. Otherwise, there have not been any significant contributions to the study of somatotopy on scalp-EEG to our knowledge.

We therefore studied SEPs on scalp-EEG during stimulation of median nerve, ulnar nerve, and all the individual digits in 3 normal volunteers and one epilepsy patient. We recorded SEPs simultaneously from a large number (32–48) of closely spaced electrodes located over the contralateral fronto-parietal region. We used dipole modeling techniques to study somatotopy of the neuronal sources underlying SEPs. Specifically, we inferred the question whether we could identify somatotopic features of hand sensory cortex, and if so, whether we could clearly separate the digits and how digit cortical representations were arranged in relation to median and ulnar nerve representations. Finally, we compared the results to those obtained from ECoG and MEG.

1.2.3.4. Somatosensory Evoked Fields on Magnetoencephalography

MEG, like scalp-EEG, is a non-invasive neurophysiological technique. As already mentioned, MEG offers the advantage of a better spatial resolution compared to scalp-EEG because skull and scalp are essentially 'transparent' to magnetic fields [162, 199]. Furthermore, the directions of highest spatial resolution, i.e. lines connecting the field extrema, are orthogonal to each other for scalp-EEG and MEG [69, 81]. This should favor MEG for the study of somatotopy of human hand somatosensory cortex as the sensory sequence follows the direction of highest sensitivity of MEG and is perpendicular to that of EEG. This better spatial resolution of MEG could be critical for exact delineation and separation of cortical digit representations.

Some authors reported on somatotopy of human hand somatosensory cortex as studied with neuromagnetic recordings [48, 268, 342]. Brenner et al. [48] recorded SEFs during stimulation of thumb and little finger and found that the magnetic field pattern during stimulation of the thumb shifted downward by approximately 2 cm compared to that during stimulation of the little finger. They found a sharp localization of the magnetic field over the active region of the cortex, which – according to these authors – was 'in contrast to the diffuse nature of potential recordings'. Okada et al. [268] measured SEFs during stimulation of the thumb, the index finger, the little finger, and the ankle. These authors applied dipole modeling and found good fits to a single equivalent dipole. The projection areas of the thumb, the index finger, the little finger, and the ankle were located at successively more medial positions along the primary somatosensory cortex. Finally, Suk et al. [342] studied thumb, index finger, and little finger and superimposed the source localizations on magnetic resonance images. All sources were located along the posterior bank of central fissure in postcentral gyrus with the little finger source located most superior and posterior, and the thumb and index finger in close proximity to each other.

However, all these studies used the steady state paradigm which does not allow investigation of the time evolution of evoked responses. Transient evoked responses, on the contrary, can be used to study the successive activation of different cortical regions followed by incoming somatosensory volleys and therefore are more appropriate to study functional organization of human somatosensory cortex. Furthermore, cortical representations were investigated only for selected digits in these studies (thumb and little finger in Brenner's study [48]; thumb, index finger, and little finger in Okada's study [268] and Suk's study [342]).

In our study, we used transient evoked responses during stimulation of all individual digits to infer functional organization of human hand somatosensory cortex and to determine whether it was possible to achieve a detailed cortical digit map by neuromagnetic recordings. Furthermore, we studied the relationship of individual cortical digit representations to the cortical regions activated during stimulation of median and ulnar nerve. Therefore, we recorded SEFs during stimulation of median nerve, ulnar nerve, and all the individual digits in 4 young healthy volunteers. We applied dipole modeling techniques to study the three-dimensional locations and time activities of the neuronal sources underlying the SEFs. Finally, we compared the results with those obtained from electric measurements on ECoG and scalp-EEG.

2. General Methodology

2.1. Neurogenesis of ECoG, Scalp-EEG, and MEG

Neuronal activity in the human brain induces ionic currents across neuronal membranes. These ionic currents generate electric fields which can be measured at the cortical surface in the electrocorticogram (ECoG) and at the scalp in the electroencephalogram (scalp-EEG). Furthermore, each electrical current induces a weak magnetic field according to Biot-Savart's law which can be measured outside the head in the magneto-encephalogram (MEG) [223]. Several important points concerning the biophysical principles underlying the neurogenesis of ECoG, scalp-EEG, and MEG have to be mentioned in the following.

ECoG, scalp-EEG, and MEG do not provide information on a local microscopic level concerning activity in individual neuronal elements, but rather measure activity in a large neuronal pool at a macroscopic level. Due to the law of superposition the electric and magnetic fields of individual neuronal elements sum up to the field of the whole neuronal population. Furthermore, the measurements of scalp-EEG and MEG, and with some restrictions also of ECoG, are performed at a distance which is large compared to the dimensions of the neuronal sources. Thus, synchronized neuronal activity of large neuronal populations is picked up in the ECoG, scalp-EEG, and MEG whereas local activity tends to cancel out [83, 88, 141, 223, 258, 337].

Furthermore it is important to consider the anatomical architecture of the cerebral cortex. Cortical pyramidal cells which are arranged in parallel columns perpendicular to the cortical surface are considered the main generators of ECoG, scalp-EEG, and MEG [223, 258]. At rest, the inner of these cell is electrically negative compared to the outside [258]. Incoming afferent impulses result in synaptic activity at the apical dendrites of these cortical pyramidal cells and cause local changes in membrane conductivity. In the case of the primary somatosensory system, afferent impulses originate usually in the ventral posterior lateral nucleus of the thalamus (VPL) and can be either excitatory or inhibitory. Whereas excitatory synaptic potentials (EPSPs) are characterized by a membrane depolarization due to a inward flow of positive ions, inhibitory synaptic potentials (IPSPs) result in a membrane hyperpolarization [258]. Thus, current is injected at the synaptic

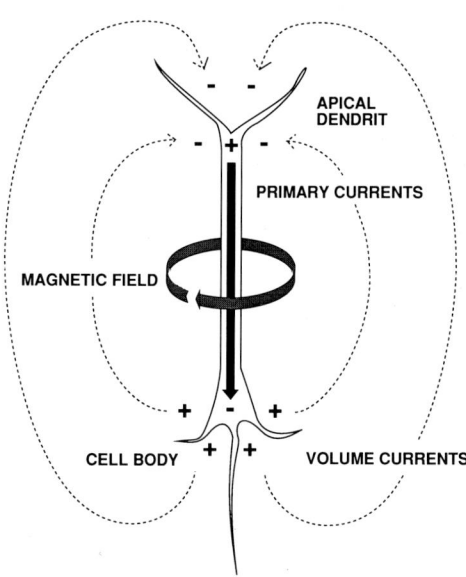

Fig. 2.1. Neurogenesis of ECoG, scalp-EEG, and MEG. Synchronized excitatory synaptic potentials (EPSPs) at the cortical pyramidal cells result in active intracellular current sources (+) and extracellular current sinks (−) at the synaptic level, and passive intracellular current sinks (−) and extracellular current sources (+) at the level of the cell bodies. This results in intracellular (primary) and extracellular (volume) currents. On ECoG and scalp-EEG, potential differences induced by volume currents can be measured. MEG, on the contrary, mainly detects magnetic fields induced by intracellular currents

level (impressed or injected current) [337]. Because there is no accumulation of charge anywhere in the medium, these impressed currents are compensated by electric currents flowing in the medium which are governed by Ohm's law (volume currents) [223, 337]. When a large population of neurons is activated simultaneously, the longitudinal components of these currents (i.e. parallel to the cell axis) add together whereas their transversal components (i.e. orthogonal to the cell axis) tend to cancel out and a laminar current in the longitudinal axis will flow [223]. These longitudinal extracellular currents can be measured at different cortical levels with specially designed laminar micro-electrodes. Thus, for both physiological and pathological processes, a regular sequence of current sources and sinks resulting in dipolar field patterns could be observed [20, 22, 23, 111, 282, 283, 284, 286, 290, 291, 292, 299, 300, 362]. Synchronized synaptic activity thus can be roughly approximated by a layer of current dipoles. In conclusion, EPSPs result in active intracellular current sources and extracellular current sinks at the synaptic level, and passive intracellular current sinks and extracellular current sources at the level of the cell bodies (Fig. 2.1). IPSPs, on the contrary, result in active intracellular current sinks and extracellular current sources at the synapses, and passive intracellular current sources and extracellular current sinks at the cell bodies. Therefore, intracellular (primary currents) and extracellular currents (volume

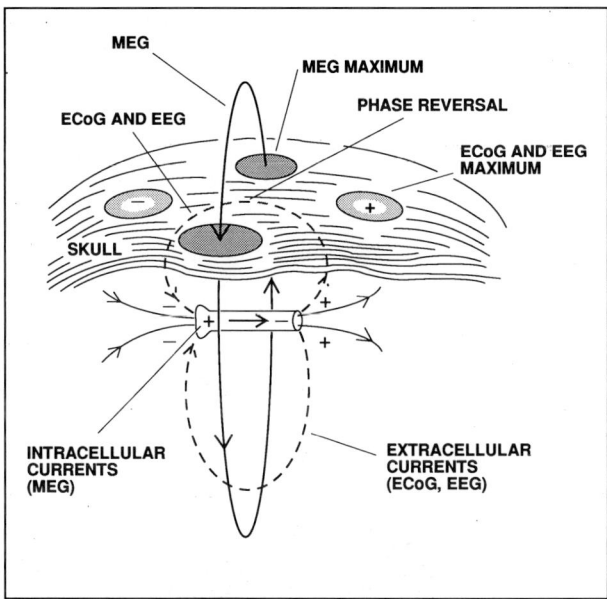

Fig. 2.2. Neurogenesis of ECoG, scalp-EEG, and MEG. On ECoG and scalp-EEG, potential differences induced by volume currents can be measured. On MEG, on the contrary, mainly magnetic fields induced by intracellular currents are detected whereas magnetic fields induced by volume currents cancel out in a spherical volume conductor. Electric and magnetic field extrema and thus the directions of highest sensitivity are orthogonal to each other. Magnetic fields are less distorted by the resistive properties of the skull and scalp which results in a better spatial resolution of MEG compared to scalp-EEG. Therefore, magnetic field maxima are about one third closer to each other than electrical potential maxima

currents) are generated. On ECoG and scalp-EEG, potential differences induced by volume currents can be measured. MEG, on the contrary, mainly detects magnetic fields induced by intracellular currents whereas the magnetic fields induced by volume currents tend to cancel out (Fig. 2.2) [150]. These issues will be further elucidated in Section 2.3.4.

The structural anatomical arrangement of cortical pyramidal cells is important for the concept of EEG and MEG generation introduced above. Due to their arrangement in parallel columns perpendicular to the cortical surface synchronized synaptic activity at these cells results in so-called open fields which can be measured throughout the medium, specifically at the surface of the brain and scalp [224, 225]. On the contrary, a different cell architecture results in another situation. Synaptic activity at cells with a radial arborization of dendrites, e.g. in the oculomotor nucleus or in the superior olive, results in volume currents which tend to cancel out not only in the tangential but also in the radial direction [223]. Thus, electric potential differences cannot be measured outside the activated neuronal population or at the surface resulting in a 'silent' EEG [208, 297].

Finally, it should be mentioned that action potentials – in contrast to synaptic potentials – do not contribute significantly to EEG and MEG generation for the following reasons. First, the area of membrane depolarization or hyperpolarization is rather small for action potentials compared to synaptic potentials. Second, the time duration of action potentials is considerably smaller than those of synaptic potentials, i.e. 1 – 2 ms for action potentials versus 10 – 250 ms for synaptic potentials [223, 258]. Thus, it can be concluded that it is synchronized synaptic activity at cortical pyramidal cells arranged in columns perpendicular to the cortical surface that is measured in the ECoG, scalp-EEG, and MEG.

2.2. Biophysical Laws of EEG and MEG

2.2.1. Forward and Inverse Problem

Both electric and magnetic fields induced by neuronal currents can be predicted because they obey physical laws. Given a set of intracerebral current sources of given location, the electric and magnetic fields generated by these neuronal elements can be predicted at the cortical surface as well as at the scalp. This problem is called the forward problem of electrophysiology [223]. The governing physical equations of this problem are the Maxwell equations in their quasistatic form [88, 137, 258, 289, 314, 337].

The inverse problem consists of calculating backwards from the electric potentials and magnetic fields actually measured at the surface unto the three-dimensional intracerebral locations and time activities of the underlying neuronal generators [83, 84, 202, 337, 338, 372]. This problem was originally formulated by Helmholtz [170]. The inverse problem has no unique solution, i.e. it is theoretically impossible to retrieve the intracerebral current sources from ECoG, scalp-EEG, and MEG measurements. Therefore, the so-called restricted inverse problem has to be formulated which means that model assumptions have to be made concerning the current sources and the volume conductor [83, 84, 202, 337, 338, 372].

2.2.2. Assumptions Concerning the Current Sources

Concerning the current sources, configuration and number of the sources have to be specified. Because the sources cannot be uniquely determined, the field pattern is often explained by so-called equivalent sources which would generate field patterns similar or 'equivalent' to those measured. The simplest equivalent source is the current dipole, the 'equivalent' to a patch of activated cortex. The dipole concept is justified for several reasons. First, as already mentioned in Section 2.1, the measured electric and magnetic fields are generated by synchronized synaptic activity at the apical dendrites of cortical pyramidal cells which are organized in columns perpendicular to the surface of the cortex and synchronized synaptic activity at these cells results in dipolar neuronal sources [223]. Second, any current source with equal number of positive and negative charges, i.e. multipoles, can be reasonably approximated by a simple dipolar source. This can be explained by the fact that the higher order terms (e.g.

quadrupolar, octupolar, etc.) fall off with distance more rapidly than the dipolar terms. Therefore, the electric and magnetic fields generated by these sources are similar to that generated by a dipolar source when the measurements are made at a distance which is large compared to the dimensions of the source – a condition that is fulfilled for both scalp-EEG and MEG, and with some restrictions also for ECoG [258]. Third, even extended sources like dipole layers or sheets approximating areas of activated cortex can be represented as simple dipolar sources without introducing significant errors in localization [89]. It has to be mentioned, however, that the spatial extent of a source cannot be determined from this models as activity arising from a dipole sheet of 1 – 2 cm in diameter cannot be distinguished from a point source [263]. Indeed, it is the center of mass that is estimated rather than the area of activated cortex. Thus, dipoles should be viewed as useful mathematical abstractions rather than as real neuronal elements [302]. In this study, we therefore chose equivalent dipoles as models of the source configuration.

In the case of ECoG and scalp-EEG, each equivalent dipole is uniquely determined by 6 parameters: 3 location parameters, and 3 orientation parameters [83, 88, 337]. In a spherical volume conductor, MEG detects only tangential dipole components and therefore it cannot be differentiated between two dipoles with identical tangential, but different radial components [150]. Thus, in the MEG only 5 dipole parameters are necessary – 3 location parameters, and 2 orientation parameters specifying the orientation of the dipole in the tangential plane, i.e. a plane orthogonal to the radius of the sphere [88, 314, 337, 369, 370].

Concerning the number of sources, most previous authors used single equivalent dipole solutions to study three-dimensional localization of the neuronal sources underlying electric and magnetic measurements. Furthermore, analysis was restricted to selected time points given by peak latencies, constancy of field distribution, or field power [83, 337, 370, 372]. However, it is agreed that the human somatosensory evoked response is generated by multiple simultaneously active brain sources producing fields that overlap both in space and time [8, 32]. Modeling activity which is generated by multiple brain areas by a single dipole can result in erroneous conclusions concerning its neurogenesis [40, 259]. We therefore used a technique that allows modeling of multiple simultaneously active brain sources over the entire time domain of the evoked response and thus could investigate its spatiotemporal structure. This modeling procedure will be presented in detail in Section 3.

2.2.3. Assumptions Concerning the Volume Conductor

Concerning the volume conductor, geometrical and conductivity properties have to be specified. Brain tissue, cerebrospinal fluid, skull, and scalp have different electric conductivities. Either realistically brain shaped or spherical volume conductors can be assumed. Brain shaped models allow for realistic modeling of head geometry and conductivity changes. These models are more accurate. Their disadvantage is that they are computationally elaborate as there exist no analytic solutions and numerical techniques, i.e. finite difference or finite element models, have to be employed [157, 242, 243, 337, 338]. Spherical models, on the contrary, only allow for modeling of concentric changes in conductivity. They are simpler as analytic solutions have been provided and inverse solutions can be obtained with reasonable computational effort on a personal computer [83, 84, 202, 337, 338, 369, 370, 372]. In this study, models using spherical volume conductors were used for the following reasons. Human primary somatosensory cortex is located in the parieto-frontal region where the surface of the scalp and brain can be reasonably approximated by a spherical model. Furthermore, computer simulation studies have shown that localizations obtained with the spherical model corresponded very closely to those derived with brain shaped models in this region of the head [157].

For the study of SEPs on ECoG, we chose a homogeneous spherical model as measurements from the cortical surface are not influenced by inhomogenities of skull and scalp. For the scalp-EEG, a 4-shell concentric spherical model with different conductivities of brain, cerebrospinal fluid, skull, and scalp was used. Whereas the homogeneous spherical model requires specification of 2 parameters (one sphere radius and one conductivity), for the 4-shell spherical model 8 parameters [4 sphere radii and 4 conductivities) have to be specified. We chose those parameters according to reports in the literature [15, 83, 136, 337]. Special considerations apply for the MEG, as the magnetic field is not affected by concentric changes in conductivity and thus a homogeneous volume conductor model can be used without introducing an error [314, 337, 338, 369, 370].

2.2.4. Practical Outline of Dipole Modeling

Once these basic assumptions concerning the source configuration and the volume conductor have been made, the electric potentials and magnetic fields generated by a given set of dipoles at known location and orientation can be calculated, i.e. the forward problem can be solved. In a spherical volume conductor, the problem consists mathematically of a Neumann problem as Poisson equations have to be solved. The solution essentially is a series of weighted Legendre polynomials and of weighted associated Legendre polynomials [83, 314, 337, 338, 369, 370].

For the solution of the inverse problem, the dipole generating a field pattern which most closely reproduces the actually measured field has to be found. This is achieved by minimizing the sum of squared differences between the dipole field and the measured field with respect to the dipole parameters, i.e. a least squares fit is performed resulting in a non-linear minimization problem. We used the simplex algorithm of Nelder and Mead to perform this minimization procedure [255, 295].

2.3. Magnetoencephalography – Basic Concepts

2.3.1. Basic Principles of the Procedure

As already mentioned in Section 2.1, synchronized neuronal currents induce very weak magnetic fields which can be measured with the new neurophysiological technique of magnetoencephalography. However, the magnetic field of the brain is extremely small. The magnetic field strength associated with evoked cortical activity is about 10^1 fT (femto Tesla) and that of the human alpha rhythm is in the order of 10^3 fT. The electrocardiogram produces a magnetic field of about 10^4 fT. On the contrary, the ambient magnetic noise in an urban environment lies in the order of 10^8 fT and therefore is considerably larger compared to the signal of interest [303, 369, 370, 371]. Thus, the investigator is confronted with the two essential problems of biomagnetism: weakness of the signal and strength of the competing environmental noise [371].

Whereas the magnetic activity of the heart was already measured in 1963 by Baule and McFee [31], the first study reporting measurement of the human magnetic alpha rhythm was published in 1968 by Cohen [68]. In this study, an induction coil magnetometer was used and the alpha rhythm only could be detected by averaging. The development of a new very sensitive generation of sensors – the SQUIDs (superconducting quantum inference devices) facilitated a major breakthrough in the investigation of the brain's magnetic field. Subsequently, MEG helped to better understand physiological as well as pathological brain activities. MEG was used to study functional organization of the primary receiving areas, i.e. the auditory cortex [160, 161, 164, 165, 270, 271, 272], the visual cortex [3, 210, 337], and the somatosensory cortex [34, 36, 42, 48, 155, 162, 166, 176, 177, 197, 200, 268, 309, 311, 346, 352, 353, 375]. Furthermore, MEG was useful to understand brain activity preceding and associated with voluntary movements. This avenue of research was pioneered by Deecke and coworkers [62, 63, 64, 90, 91, 92, 93]. Besides these mainly neurophysiologically oriented studies, the main clinical application of MEG has been human epilepsy. MEG has helped to get more accurate localization estimates of the epileptogenic lesion in patients with focal seizures [21, 27, 28, 29, 41, 44, 301, 305, 315, 316, 317, 330]. This may be of clinical relevance for the presurgical evaluation of epilepsy patients especially if selective surgical techniques are used [366, 367]. Furthermore, MEG contributed to our understanding of the pathophysiology of human epilepsy as it helped to study the mechanisms of neocortical spike propagation [344] and slow magnetic field shifts associated with epileptiform discharges [24, 25]. Several reviews concerning various applications of MEG have been published [35, 43, 163, 303, 306, 318].

2.3.2. Instrumentation

Technical aspects of instrumentation have been extensively discussed in several articles and are beyond the scope of this study [196, 303, 304, 369,

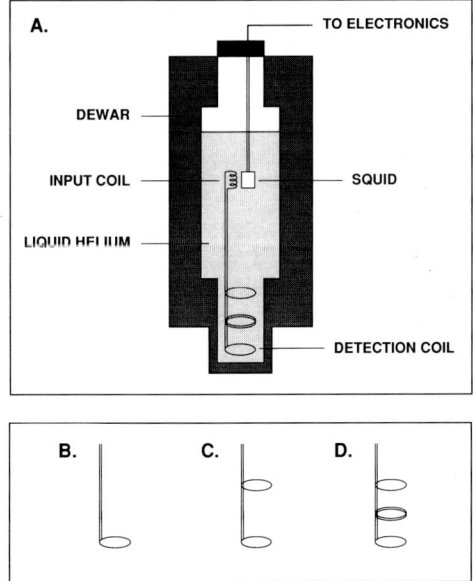

Fig. 2.3. A. Magnetometer in schematic form. The magnetic field is sensed by a detection coil forming a closed superconducting circuit (flux transformer) with an input coil in close proximity to the SQUID. The electric current in the flux transformer impresses a field to which the SQUID responds. The response of the SQUID is monitored by an electric circuit including a sensitive preamplifier (SQUID electronics). The SQUID and its associated superconducting components are kept at a temperature of 4.2K by immersion in a bath of liquid helium within a cryogenic vessel (dewar). **B.** Magnetometer. **C.** First-order gradiometer. **D.** Second-order gradiometer. Whereas first order gradiometers consist of two coils which are connected in series, second order gradiometers consist of four coils. Gradiometers thus respond only to magnetic field gradients, but are insensitive to homogeneous fields generated by distant sources. Second-order gradiometers have a lesser overall sensitivity compared to first-order gradiometers, but this reduction in sensitivity corresponds to a higher degree of spatial discrimination. Thus, second-order gradiometers are more appropriate to study cortical sources

371, 392]. We therefore only mention some aspects especially relevant to the application of MEG in studying somatosensory evoked magnetic fields.

The core of each biomagnetic recording device is the SQUID (superconducting quantum inference device). SQUIDs are based on superconductivity, the underlying physical principle is the Josephson effect [189]. Fig. 2.3A shows a SQUID system in schematic form. The brain's magnetic field is sensed by a detection coil which should be placed as close as possible to the field source. This detection coil forms a closed superconducting circuit with an input coil in close proximity to the SQUID. This closed circuit is known as the flux transformer transforming magnetic flux into electric current for the following reason. One property of a closed superconducting loop is that if a magnetic field is applied anywhere within the loop, the superconducting electrons flow through the wire so that their current produces a field that maintains the net magnetic flux in the loop (product of field and area) invariant. Consequently, if a magnetic field passes through the loops of the detection coil, current passes around the entire circuit, and the portion flowing in the input coil impresses a field on the SQUID to which the SQUID responds [371]. The response of the SQUID is monitored by an electric circuit including a sensitive preamplifier (SQUID electronics). Thus, the SQUID can be regarded as a highly sensitive current to voltage amplifier with an output linearly related to the instantaneous value of the magnetic field at the detection coil [306, 371]. The latest generation of very sensitive SQUIDs – the so-called direct current biased SQUIDs (dc-SQUIDs) – are fabricated using highly sophisticated photolithographic techniques [57, 303] and consist of two arms of superconducting film interrupted by two Josephson junctions through which direct current is passed [303, 371]. As SQUIDs and its associated components are based on superconductivity, they have to be kept at a temperature of 4.2K by immersion in a bath of liquid helium within a cryogenic vessel (dewar) [303, 304, 369, 371].

The geometry of the detection coils can be modified in order to achieve partial insensitivity to environmental magnetic noise. Figs. 2.3B – 2.3D show a simple magnetometer and two of the most commonly used types of detection coils – a first-order and a second-order gradiometer. Whereas first order gradiometers consist of two coils which are connected in series, second order gradiometers consist of four coils. Gradiometers thus respond only to magnetic field gradients, but are insensitive to homogeneous fields generated by distant sources. Second-order gradiometers have a lesser overall sensitivity compared to first-order gradiometers, but this reduction in sensitivity corresponds to a higher degree of spatial discrimination [303]. This means that second-order gradiometers are most useful to study cortical activity as somatosensory evoked magnetic fields. Furthermore, the size of the detection coil determines sensitivity and spatial resolution. High sensitivity implies a large detection coil, whereas good spatial resolution requires a small coil. However, no practical benefit exists in using detection coils with a diameter smaller than the distance between the coil and the source [123] which can be assumed as 2 cm for brain measurements [162].

Whereas initially only single channel systems were available, multichannel systems consisting with 4, 5, 7, 9, and recently with 24 and 37 channels within one dewar are now used in several laboratories [151, 156, 178, 195, 209, 328]. The advantage of multichannel systems is that the magnetometer needs not to be moved to successive measurement sites and recordings can be performed in a 'single shot'. This is especially useful for the investigation of spontaneous brain activity, e.g. epileptiform discharges, and of higher cognitive functions. For these studies, multichannel systems should allow a major break-through in biomagnetism. We used a dual 7-channel magnetometer system – a so-called gemini system – with second-order gradiometers for our measurements.

A further line of development in biomagnetism may include completely different sensing units consisting of planar gradiometers microfabricated onto silicon chips, and directly integrated to the dc-SQUID within the same chip [52, 58, 205] which could provide large multi-channel systems.

2.3.3. Magnetic Shielding

A further problem for neuromagnetism is the interference of ambient magnetic noise which is always present in a hospital environment, especially in the low frequency range. Strong sources of magnetic noise include various electronic equipments, moving large ferromagnetic objects and building vibrations. Furthermore, radiofrequency noise as from communication systems may interfere with biomagnetic measurements [196, 371]. Therefore, high resolution measurements are possible only inside a magnetically shielded room which provides both low and radiofrequency electric and magnetic shielding [196, 371]. Initially, high quality shielded rooms with at least three layers of mu-metall layers were used [203, 235]. Recently, a wall construction with two layers of mu-metal and at least one layer of aluminum has become popular. This type of room is less expensive, provides adequate shielding [53] and was used in our study.

2.3.4. Differences Between Scalp-EEG and MEG

Although EEG and MEG are generated by the same neurophysiological process, namely synchronized synaptic potentials at cortical pyramidal cells, there are some important differences concerning the neurogenesis of EEG and MEG. EEG and MEG yield both complementary and confirmatory information to each other [69, 81].

In contrast to electric fields, magnetic fields are less distorted by the resistive properties of the skull and scalp [69, 150]. Some authors have even suggested that the skull and scalp are virtually transparent to magnetic fields [162, 198, 199]. This results in a better spatial resolution of MEG compared to scalp-EEG and magnetic field maxima are about one third closer to each other than electrical potential maxima (Figs. 2.2 and 2.4C) [69]. Electric and magnetic fields are oriented perpendicular to each other. Thus, the directions of highest sensitivity – usually the direction between the field maxima – are orthogonal to each other [69, 162] (Figs. 2.2 and 2.4C).

Whereas scalp-EEG is sensitive to both tangential and radial components of a current source in a spherical volume conductor, MEG detects only the tangential components (compare Figs. 2.4A, 2.4B, and 2.4C). Thus, MEG selectively measures the activity in the sulci (Fig. 2.4C), whereas scalp-EEG measures both activity in the sulci and at the top of the cortical gyri (Figs. 2.4A, 2.4B, and 2.4C). Thus, the combined use of scalp-EEG and MEG can be used to resolve ambiguities as shown in Fig. 2.4. An electric field with one electropositivity and one electronegativity, respectively, could be generated either by two radial dipoles representing distributed activity (Fig. 2.4B), or by one tangential dipole representing focal activity (Fig. 2.4C). From electric measurements alone no definitive decision can be made which situation corresponds to reality. MEG can resolve this ambiguity as two radial dipoles would not produce a measurable

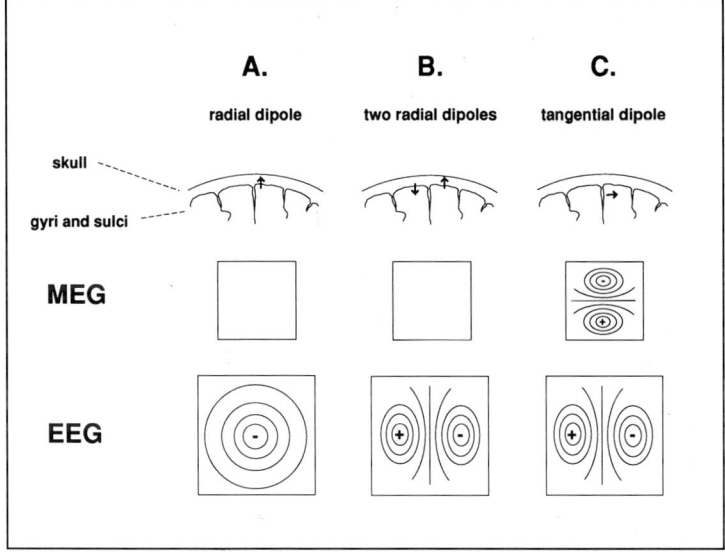

Fig. 2.4. Differences between scalp-EEG and MEG. **A.** Magnetic and electric field generated by one radial dipole. **B.** Magnetic and electric field generated by two radial dipoles. **C.** Magnetic and electric field generated by one tangential dipole. Scalp-EEG measures both radial activity at the top of the cortical gyri (A, B) and tangential activity in the sulci (C). On the contrary, MEG selectively measures tangential activity in the sulci (C) and is blind to radial activity (A, B). Electric and magnetic fields are orthogonal to each other (C). Thus, scalp-EEG and MEG yield both complementary and confirmatory information. An electric field with one electropositivity and one electronegativity, respectively, could be generated either by two radial dipoles representing distributed activity (B), or by one tangential dipole representing focal activity (C). From electric measurements alone no definitive decision can be made which situation corresponds to reality. MEG can resolve this ambiguity as two radial dipoles would not produce a measurable magnetic signal (B), whereas a tangential dipole would generate a magnetic field perpendicular to the electric field (C)

magnetic signal (Fig. 2.4B), whereas a tangential dipole would generate a magnetic field perpendicular to the electric field (Fig. 2.4C).

Both scalp-EEG and MEG measure the summated synaptic potentials of dendritic cortical synapses [223]. Whereas scalp-EEG is sensitive to extracellular volume currents produced by these synaptic potentials, MEG is primarily sensitive to intracellular currents associated with these synaptic potentials [23, 26, 265, 266, 267]. This is due to the fact that the field components generated by volume currents tend to cancel out in a spherical volume conductor (Figs. 2.1 and 2.2) [70].

Generally, the decay of the magnetic fields as a function of distance is stronger than for the electric fields [162]. MEG thus is more sensitive to superficial cortical activity which was the main interest of the present study. Another advantage of MEG – especially for the measurement of low-frequency activity – is that it is independent of artifacts generated at the electrode-skin interface. Furthermore, in MEG no reference point is needed. This is in contrast to scalp-EEG, where an active reference can lead to serious difficulties [103, 355].

2.4. Cortical Stimulations

In neurosurgical patients undergoing epilepsy surgery, surgery of tumors or arteriovenous malformations (AVMs), pre-resection assessment requires knowledge not only what to take out, but also what to leave in [218]. Localization of motor, sensory, and language-related cortex usually has been determined by electrical stimulations [219, 228, 229, 261, 262, 278, 279, 280]. Concerning the stimulation procedure, we used a protocol that was adapted from the Cleveland Clinic Foundation and has been extensively reviewed by Lüders et al. [228] and Lesser et al. [218].

2.4.1. Subdural Grid Electrodes

The electrodes are made of platinum iridium with a diameter of 6 mm and a spacing of 10 mm center-to-center. These electrodes are embedded in sheets of silicone rubber. There exist strips with 4 – 8 electrodes, and plates with 20, 32, and 64 electrodes. Whereas strips can be placed through burr-holes, the placement of the girds requires a craniotomy. Because it is necessary for the cables to exit the skull from these grids, there is always the risk of infection, and appropriate precautions have to be taken. Therefore, the cables are tunneled under the skin for a distance before exiting the scalp. Prophylactic antibiotics are administered.

Further risks of grids include pressure effects and development of subdural hematoma. Grids represent a foreign body and can exert pressure onto the underlying brain tissue. Thus, patients may show transient focal neurologic deficits after insertion [218]. Furthermore, grid implantation may increase intracranial volume up to 9 cc which has to be taken into account in patients with increased intracranial pressure [152]. Therefore, patients receive a course of dexamethasone and the implantation is performed on a Thursday allowing the brain to adapt over the weekend to the grids until stimulations are begun [218, 343]. Furthermore, for the first days after grid placement intracranial pressure is monitored [218, 343].

2.4.2. Stimulus Parameters

Stimuli were delivered by a Grass S88 stimulator via two SIU7 stimulus isolation units. The following stimulus parameters have to be considered: 1. stimulus polarity, 2. single pulse versus repetitive stimulation, 3. stimulus pulse duration, 4. stimulus frequency, 5. stimulus duration, 6. stimulus intensity, and 7. intervals between successive stimulations.

1. Stimulus Polarity. It has been estimated that unidirectional currents of 2.5 Coulomb can produce damage to the brain [234]. Therefore, we used stimuli of alternating polarity resulting in an equal number of charges passing in both directions. Furthermore, the release of ions at the metal/brain interface thus is minimized [221].

2. Single Pulse Versus Repetitive Stimulation. We used repetitive stimulation for the initial testing of each electrode as this results in lower thresholds due to temporal facilitation [228, 307]. If positive motor effects occurred with repetitive stimulation, we confirmed these findings by one-to-one motor effects elicited by single pulse stimuli at maximum intensity, i.e. 15 mA.

3. Stimulus Pulse Duration. The stimulus pulse duration was 0.3 ms. Increasing stimulus pulse duration results in a considerable increase in effectiveness, but also lowers the afterdischarge threshold [228].

4. Stimulus Frequency. It has been shown that there exists an inverse logarithmic linear relationship between stimulus frequency and stimulus intensity [220]. In accordance with Lüders et al., we used 50 Hz trains of stimuli which should produce maximum functional alteration at any given stimulus intensity [228].

5. Stimulus Duration. We used train durations of 5 s. If these stimulus durations resulted in low afterdischarge thresholds, we shortened stimulus durations to 2 s. It should be mentioned that with 2 s stimulus duration a meaningful assessment of higher cortical functions is difficult to achieve [228]. Actually, for testing higher cortical functions we sometimes used stimulus durations of up to 10 s. In agreement with Lüders et al. [228], during longer stimulus durations effects were observed to diminish after 5 – 10 s. This could be explained either by an activation of alternative pathways [227] or by an adaption with hyper- or depolarization [228].

6. Stimulus Intensity. We started stimulations with 1 mA, and successively increased stimulus intensities in steps of 0.5 or 1 mA. It has to be stressed that both positive effects as well as afterdischarge thresholds change from electrode to electrode and for an individual electrode from day to day [218, 228]. This slow and gradual increase in stimulus intensity therefore has to be performed on each individual electrode every time it is tested. Otherwise afterdischarges leading to seizures or unexpected violent motor effects could be induced [228].

7. Intervals Between Successive Stimulations. We used 20 s time intervals between successive stimulations which can be considered as adequate [218].

2.4.3. Neurophysiological Effects of Cortical Stimulations

On a microscopic level, electrical stimulation of the cortex results in synchronized de- or hyperpolarization of a large cortical neuronal pool. On a behavioral level, the effect of cortical stimulations depends on whether primary efferent or afferent cortical areas (e.g. motor, somatosensory, visual, or auditory) or association areas are stimulated [228].

Stimulation of primary cortex results in so-called positive symptoms, e.g. muscle twitches during stimulation of motor cortex, paresthesias during stimulation of somatosensory cortex, and simple visual or auditory hallucinations during stimulation of primary visual or auditory cortex, respectively [228]. In this context, it should be mentioned that cortical stimulations are highly non-physiological as they produce synchronized activation of all cortical cells in a given neuronal pool. Thus, they cannot reproduce the complex spatiotemporal patterns of excitation and inhibition which can be observed in different cortical layers during physiological as well as pathological processes [20, 22, 23, 24, 25, 26, 111, 126, 127, 128, 129, 282, 283, 285, 290, 291, 292]. This is in line with the observation that the behavioral effects are non-physiologic, e.g. paresthesias, unusual movements and usually produce inhibitory effects [228].

Stimulation of association areas produces negative effects, and thus the patient at rest is unaware that the brain has been stimulated. However, when complex actions are preformed these cannot be continued during stimulation [228]. An example of a negative effect is the negative motor response (NMR) elucidated during stimulation of pre-motor cortex. During stimulation of these brain areas, patients failed to continue rapidly alternating complex movements of the eyes, tongue, or fingers. The negative effects can be explained by the fact that input/output characteristics of association areas are even more complex than in primary cortex and therefore cannot be reproduced by cortical stimulations. Furthermore, inhibitory effects seem to predominate in these areas thus protecting the brain from non-physiological outputs [228].

2.4.4. General Testing Procedure

Cortical stimulations can be performed by stimulation of two adjacent electrodes. However, this produces sometimes ambiguities regarding which electrode a sign or symptom was localized to [218]. Thus, following the recommendations of Lesser et al. [218], we stimulated between an 'active' electrode and a 'reference electrode'. Ideally, such a reference electrode is located along the periphery of the plate in the anterior inferior portion of the temporal lobe which is planned to be resected. From a functional viewpoint, stimulation of this electrode is not allowed to produce positive or negative effects or afterdischarges. After a reference electrode has been identified each electrode is checked for positive symptoms. As already mentioned stimuli consisted of 0.3 ms pulses 50 Hz trains with alternating polarity and 5 s duration. Stimulus intensities were increased at 1 mA increments till either positive effects, afterdischarges, or the maximum stimulus intensity of 15 mA was reached. Subsequently, this electrode was checked for negative symptoms at 15 mA or at 0.5 mA below afterdischarge threshold. Reading aloud served as an initial test for negative symptoms and was followed by special tests tailored to the expected effects in this area, e.g. testing for negative motor symptoms in pre-motor cortex.

3. Spatiotemporal Modeling on ECoG, Scalp-EEG and MEG

3.1. Motivation

One of the major problems in clinical neurophysiology has been the investigation of physiological as well as pathological processes involving multiple simultaneously active brain regions using electrical and magnetic measurements [21, 29, 32, 34, 40, 41, 42, 138, 259, 319, 320, 321, 322, 323, 324, 325, 326, 379]. In this context, it is important to recognize that electric potentials or magnetic fields measured at the scalp do not merely reflect activity of neuronal populations immediately underlying the measurement site. Actually, many different and even remote neuronal populations, which are active simultaneously, contribute to scalp electric and magnetic recordings [258, 259]. Thus, the application of methods which allow modeling of multiple simultaneously active brain regions is warranted. We used spatiotemporal modeling to deal with this problem and applied a combined approach of principal component analysis (PCA) and multiple dipole modeling which will be outlined in the following.

3.2. Simulation Study

We will demonstrate the basic principles of the spatiotemporal modeling in a simple simulation experiment on scalp-EEG. Details of this simulation experiment have been presented in a separate article [40]. It should be noted that the conclusions of this experiment equally apply to ECoG, scalp-EEG, and MEG without any loss of generality.

We placed current dipoles at different locations inside a spherical volume conductor and calculated the potentials at 49 electrode sites on the surface of the sphere thus simulating a situation that could arise in the scalp-EEG. The volume conductor was assumed to be a 4-shell spherical model with different conductivities for brain, cerebrospinal fluid, skull, and scalp. The outer radius of the sphere was assumed as 8 cm approximating the adult human cranium. The radii for brain, cerebrospinal fluid, and skull were assumed as 6.7 cm, 6.9 cm, and 7.6 cm. The conductivities were set to 0.33 $(\Omega m)^{-1}$ for the brain tissue, to 0.0042 $(\Omega m)^{-1}$ for the cerebrospinal fluid, to

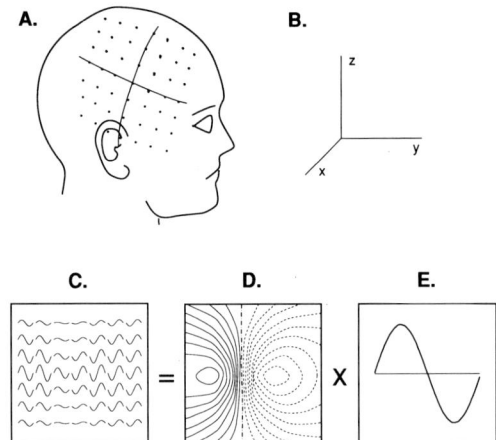

Fig. 3.1. Simulation study – single dipole. We placed current dipoles inside a spherical volume conductor and calculated the potentials at 49 electrode sites on the surface of the sphere. **A.** Schematic representation of the electrode sites on the scalp. **B.** Coordinate system. **C.** Simulated scalp potentials generated by a tangential dipole (dipole location x = 65 mm; y = -10 mm; z = 0; time activity = 2 Hz). These potentials are uniquely determined by a spatial potential pattern (D) and a time pattern (E). The spatial potential pattern is shown as an isopotential map (10% isocontour lines)

1.0 $(\Omega m)^{-1}$ for the skull, and to 0.33 $(\Omega m)^{-1}$ for the scalp. The values for the radii and for the conductivities were chosen according to values in the literature [15, 83, 84, 136, 258, 337, 338]. A rectangular 7 times 7 matrix of 49 equally spaced electrodes (interelectrode distance = 3 cm) was designed at the surface of the sphere (Fig. 3.1A). In order to simulate potentials at these electrodes, which were assumed to reflect a unipolar recording with reference at infinite distance, current dipoles were fixed at a given location inside the cranium and their activity was varied over time. The forward

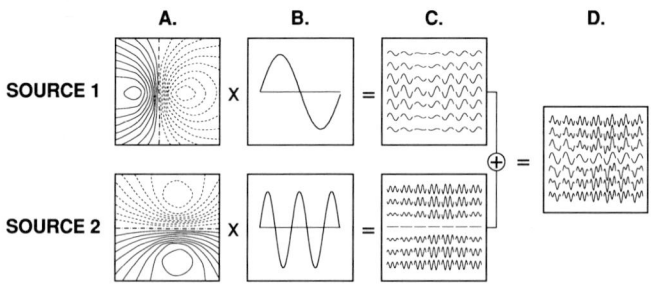

Fig. 3.2. Simulation study – multiple dipoles which overlap in space and time. **A.** Spatial potential patterns shown as isopotential maps. **B.** Time activities were set to 2 Hz and 5 Hz, respectively. **C.** The spatiotemporal patterns of the individual dipoles can be calculated by multiplying the spatial patterns (A) with the time activities (B). **D.** According to the law of superposition, the spatiotemporal patterns of the individual dipoles sum up to form the combined spatiotemporal pattern of the two dipoles which is equivalent to the potentials actually measured at the scalp

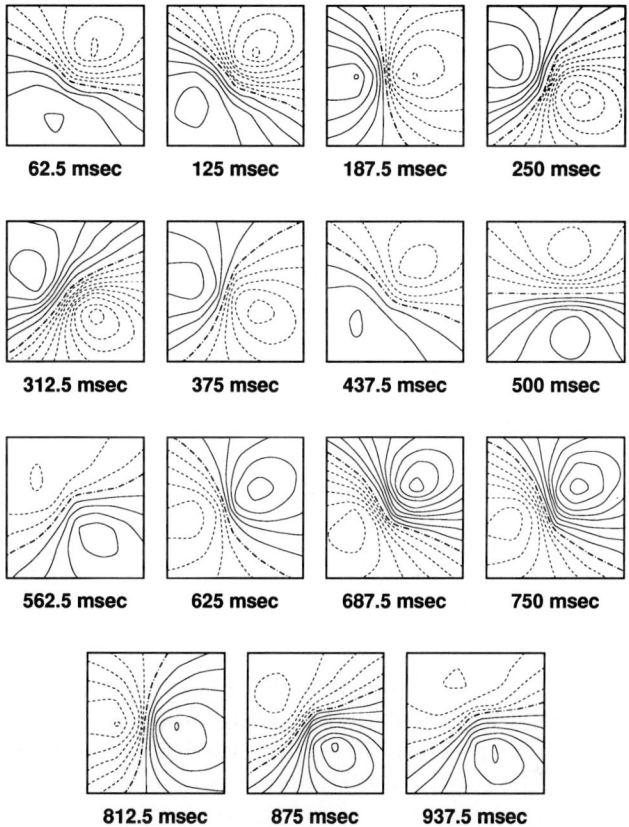

Fig. 3.3. Isopotential maps for the scalp-EEG waveforms generated by two simultaneously active sources (spatial patterns, time patterns, and spatiotemporal patterns of the individual and combined sources are shown in Fig. 3.2). The isopotential maps are no longer constant over time, but instead show different configurations for successive time points. The contribution of the two sources to the maps varies both over space and time. Thus, the isopotential maps can easily be misinterpreted. Isopotential maps: 10% isocontour lines; each map scaled to maximum

solution was calculated according to standard formulas [83, 84, 337, 338] (see also Section 2.2.1). The head coordinate system was assumed as follows: origin: center of the sphere; x-axis: pointing towards the right; y-axis: pointing anterior; z-axis: pointing upward (Fig. 3.1B).

Fig. 3.1C shows the potentials at the 49 electrodes generated by a tangential dipole (dipole location x = 65 mm; y = -10 mm; z = 0; time activity = 2 Hz). These potentials are uniquely determined by a spatial potential pattern (Fig. 3.1D) and a time pattern (Fig. 3.1E). In Fig. 3.1D the spatial pattern is depicted as an isopotential map. Note, that the spatial pattern remains constant over time and reflects the anatomically fixed location of the dipole. A rough estimate of the dipole location can be achieved by visual inspection of the map without any mathematical modeling. A line connecting the potential maxima indicates the tangential dipole axis and the polarity reversal point its location. The time pattern reflects the time varying activity of the dipole (Fig. 3.1E).

However, the situation becomes more complicated, when multiple neuronal populations are active simultaneously. In order to demonstrate this, we implanted a second dipole with an approximately orthogonal orientation to the first dipole into the sphere (dipole location x = 60 mm; y = 10 mm; z = 0; time activity = 5 Hz). The potentials generated by the two dipoles now overlap both in space and time. Thus, the potentials at each electrode site consist of time and space varying contributions from two dipoles. Fig. 3.2 shows the spatial patterns (Fig. 3.2A), the time activities (Fig. 3.2B), and the resulting spatiotemporal patterns (Fig. 3.2C) at the 49 electrodes for the two dipoles separately. Due to the law of superposition, the spatiotemporal patterns of the individual dipoles simply sum up to form the combined spatiotemporal pattern (Fig. 3.2D) of the two dipoles which is equivalent to the potentials actually measured at the scalp.

Interpretation of the isopotential maps of the simulated 'raw data' becomes more complicated compared to the situation when only one dipole is active. The isopotential maps are no longer constant over time, but instead show different configurations for successive time points (Fig. 3.3). The contribution of the two sources to the maps varies both over space and time. Thus, the isopotential maps could easily be misinterpreted. This especially occurs when single dipoles are fit to the data at successive time points as it is done in the so-called 'moving dipole models' [79]. In our special situation, this single dipole would move back and forth between the actual locations of the two dipoles and its orientation would be a mixture of the actual two dipole orientations. Thus, modeling activity that is generated by multiple brain regions with a single dipole can yield misleading results [40, 259].

3.3. Basic Goals of Spatiotemporal Modeling

From this simple example, the basic goals of spatiotemporal modeling can be deduced. In short, we have to work our way backwards from the potentials measured at the surface (Fig. 3.2D), first to the spatiotemporal contributions of the individual sources (Fig. 3.2C), and finally to their spatial (Fig. 3.2A) and temporal patterns (Fig. 3.2B). Thus, the following steps have to be performed. First, the number of sources underlying a given neurophysiological process has to be determined. Second, the electric potential contribution (scalp-EEG and ECoG) and magnetic field contribution (MEG) of each source at each measurement site has to be clearly identified. Finally, the anatomical locations and time activities of the sources have to be determined. It is only the last step that dwells into the inverse problem.

To solve these problems, we applied a combined approach of the multivariate statistical technique of PCA and of biophysical modeling. We used PCA to determine the number of underlying sources and to identify their potential and field contributions. Subsequently, we incorporated the results of PCA into a biophysical model which was used to solve the inverse problem, i.e. to estimate the anatomical locations and time activities of the neuronal sources. Furthermore, the biophysical model was used to reassess the validity of the first two steps of the analysis, i.e. the number of sources active and their spatiotemporal potential or field contributions.

3.4. Principal Component Analysis

3.4.1. Introduction

The application of PCA was introduced to neurophysiological research by Donchin [116]. Originally, PCA was mainly used to study the influence of experimental conditions on event-related potentials or averaged evoked responses [140, 247, 308, 363, 376]. In order to investigate the simultaneous activity of multiple brain regions in a single subject, however, different statistical model assumptions have to be made and were originally proposed by Kavanagh et al. [201]. In this context, the goal of PCA is to explain the variance of the ECoG, scalp-EEG, or MEG signals which are measured at many distinct sites by a much smaller number of underlying 'basic waveforms' or principal components which ideally should correspond to actual brain sources. Subsequently, several investigators used this approach. Wood et al. [377] reported preliminary results on somatosensory evoked responses on scalp-EEG. Maier et al. [236] applied a similar technique to visually evoked potentials. Achim et al. [2] and Baumgartner et al. [40] performed detailed simulation studies. Finally, Baumgartner et al. used this technique to study somatosensory evoked magnetic fields [34] and the epileptic spike complex on MEG [41].

3.4.2. Methods

PCA is a data reduction technique [112, 167, 249]. As already mentioned, the goal of PCA therefore is to explain the variance of the scalp-EEG, ECoG, or MEG signals at many measurement sites by a much smaller number of underlying 'basic waveforms' or principal components. The data are thus represented in a space of much smaller dimension, the so-called 'common factor space'. The principal components are derived as those linear combinations of the original data time series, which successively explain a maximum of the system variance and are orthogonal to each other. The principal components are determined by calculating the eigenvectors of the system covariance matrix. The spatial weighting factors of these linear combinations are called factor loadings and the actual time varying values of the basic waveforms are called factor scores [112, 167, 249].

The original data can be reconstructed by multiplying the spatial pattern of factor loadings which remain constant over time with the factor scores which vary over time. On physiological terms, the factor loadings can be interpreted as a potential or magnetic field distribution generated by a given neuronal population anatomically fixed in space, whereas the factor scores correspond to the time activity of this neuronal population. The linear model is justified by the physical law of superposition [258]. The resulting model, therefore, can be written as follows:

$$y_{ij} = \sum_{k=1}^{nsrc} b_{jk} * a_{ki}$$

with y_{ij} = electric potential or magnetic field at measurement site j at time point i, b_{jk} = factor loadings of principal component k at measurement site j, a_{ki} = factor scores of principal component k at time point i;

> i = 1,..., ntim: number of time points
> j = 1,..., nmea: number of measurement sites
> k = 1,..., nsrc: number of principal components

The eigenvalues and eigenvectors of the covariance matrix were calculated by tridiagonalization and the QL-algorithm [295]. The number of significant principal components was determined by two independent criteria. First, we used the 'eigenvalue greater than one criterion' which is a commonly used method [112]. Second, we developed a 'variance criterion' as follows. We estimated the signal noise by determining run-to-run variability which was calculated by dividing the sum of squared differences between two successive runs by the sum of squares of the first run. If the variance explained by a given principal component was smaller than this run-to-run variability, this particular principal component was considered as insignificant.

It has to be mentioned, however, that the results of PCA are not unique. Once the common factor space has been determined by principal component extraction, it is in no way defined by a particular coordinate system. Thus, an infinite number of rotations is possible from one coordinate system to another without any effect on the adequacy of the solution [167]. Therefore, a number of rotations of the principal components have been used in order to improve the interpretability of the results of PCA. The common goal of all these rotations is to achieve a 'simple structure' of the spatial or temporal patterns of the underlying hypothetical brain sources [40, 112, 140, 167, 247, 363, 376, 377]. These rotations will not further be elucidated here, for a detailed discussion the reader is referred to a separate publication [40].

3.5. Combination of Principal Component Analysis and Biophysical Modeling

As already mentioned, PCA is a purely statistical technique with no relationship to physiological reality. Therefore, it seems desirable to relate the results of PCA to a biophysical model. One possible approach was recently reported by Maier et al. [236] in a study of visual evoked responses and consists of rotation of the principal components to a physiological, i.e.

dipolar structure. This method seems very straightforward and attractive from a physiological viewpoint, but imposes mathematical and computational difficulties. The procedure essentially consists of fitting multiple dipoles in the common factor space. However, the problem of fitting multiple dipoles at one time point has been shown to be a computationally unstable procedure [83]. Therefore, this approach has to be limited to experiments where a detailed a-priori knowledge of the actual dipole parameters exists, as in Maier's study of visual evoked responses.

Another approach was proposed by Wood et al. [377] and consisted of fitting single dipole to the individual principal component loadings representing spatial potential or field distributions. It should be noted, that these spatial patterns remain constant over time and thus this procedure is independent of selection of a special point in time. The basic assumption of this approach is that the principal components represent the distinct brain sources reasonably well. However, this assumption is no physiological necessity. In this context, the orthogonality assumption of PCA becomes a problem, especially when the activity of the sources is strongly correlated [360]. The advantage of this approach is its computational stability.

We used the technique proposed by Wood et al. [377] in an intermediate step between PCA and multiple dipole modeling, i.e. to obtain reasonable start values for the non-linear minimization procedure in the multiple dipole model.

3.6. Multiple Dipole Modeling

3.6.1. Introduction

Biophysical models have to be applied to solve the inverse problem in neurophysiology, i. e. to calculate backwards from the electric potentials or magnetic fields measured at the scalp (scalp-EEG and MEG) or on the cortical surface (ECoG) unto the three-dimensional intracerebral locations and time activities of the underlying neuronal sources. As already mentioned in Section 2.2, dipole models have become very popular to solve the inverse problem both because of their physiological reasonability and because of their mathematical simplicity. Most previous investigators, however, used only single dipole models and restricted their analysis to selected time points or intervals given by peaks [83, 176, 177, 200, 337, 338, 346, 352, 372], maximum field power [84, 214] or constancy of scalp potential distribution [202, 379]. Other investigators used so-called 'moving dipole solutions' where single dipole fits are performed at successive time points, and the dipoles are allowed to move in the brain [79, 330]. The basic assumption of these models is that there exist time instants when predominately only one brain source is active and that sequential neuronal activation can be modeled by moving dipoles. However, as our simulation experiment demonstrates, modeling activity generated by multiple brain areas with a single dipole can lead to erroneous results (Section 3.2). Specifically, time varying isopotential maps are not necessarily generated by sequential activation of different neuronal populations, but rather can arise from anatomically fixed sources with time varying activities (Figs. 3.2 and 3.3). These findings are in agreement with the results reported by Nunez [259].

Multiple dipole modeling is another method to solve the inverse problem. This approach exploits the entire spatiotemporal information of brain activity and allows modeling of multiple simultaneously active brain sources. Similar to the PCA approach, the data first are decomposed into spatiotemporal contributions of the individual brain sources. Furthermore, the actual three-dimensional intracerebral locations and time activities of the sources can be estimated as the inverse problem is solved.

The modeling assumptions underlying this approach are that the dipoles are fixed in space and that their activities vary over time. The basic ideas of this concept were initially introduced in neurophysiological research by Scherg et al. [319, 324, 325, 326]. The model assumption may be justified from neuroanatomical considerations as neuronal populations do not move around in the brain, but rather are anatomically fixed and vary in activity over time. Initially, Scherg et al. [319, 324, 325, 326] restricted the time activity of the sources to waveforms determined by 6 parameters (2 peak, 2 onset latencies and 2 offset latencies) and used spline interpolation to calculate the time activity once these parameters had been estimated. Furthermore, dipole locations were restricted to a two-dimensional plane. This approach requires detailed a prior knowledge concerning source location and activity. Subsequently, the multiple dipole model was generalized independently by these authors and several other investigators [2, 21, 32, 34, 40, 41, 88, 215, 236, 320, 321, 322, 323]. The advantages and shortcomings of this approach have been investigated in computer simulation studies [2, 40]. Furthermore, the model was applied in studies of the auditory cortex [320, 322, 323], the visual cortex [236], the somatosensory cortex [32, 34, 41, 42], and to the study of the human epileptic spike complex [21, 41].

This approach seems especially appropriate for modeling the human somatosensory evoked response which is believed to be generated by multiple simultaneously active brain sources [8, 346, 378]. In the present study, we therefore applied spatiotemporal modeling to somatosensory evoked responses recorded on ECoG, scalp-EEG, and MEG.

3.6.2. Methods

3.6.2.1. Forward Problem

The forward solution was obtained in analogy to the simulation study presented in Section 3.2. In a first step, the spatial potential or magnetic field pattern generated by a dipole of known location and orientation has to be calculated (Fig. 3.2A). We applied standard formulas from the literature providing the electric and magnetic fields generated by a single equivalent dipole outside a spherical volume conductor [83, 84, 337, 338, 369, 370]. In the next step, these spatial patterns were multiplied with time varying source activities (Fig. 3.2B) in order to obtain the spatiotemporal contributions of the individual sources (Fig. 3.2C). It should be noted that the dipole location and orientation parameters representing the spatial source pattern remained constant over time. Finally, the spatiotemporal patterns of the individual sources were summed up to form the combined spatiotemporal pattern due to the law of superposition (Fig. 3.2D). This combined spatiotemporal pattern should reproduce the data as closely as possible.

3.6.2.2. Inverse Problem

Thus, the goal of the modeling procedure was to find that combination of spatial and time patterns, i.e. spatiotemporal patterns which explained a maximum amount of the data variance. Variance accounted for was calculated as the sum of squared differences between the data and the model divided by the sum of squares of the data. It is this sum of squared differences that has to be minimized in the multiple dipole model.

The minimization procedure involves a linear part for identifying the time activities and a non-linear part for identifying the source location parameters. We applied standard linear regression techniques for linear minimization [118]. The simplex algorithm of Nelder and Mead was used to solve the non-linear minimization problem [255, 295]. For this non-linear minimization, initial guesses concerning the dipole locations and orientations, i.e. so-called start values, have to be provided. An appropriate selection of start values is critical for the success of the procedure, i.e. to achieve correct location and time activity estimates. These start values can be derived from a-priori physiological knowledge. However, this a-priori knowledge is not always available. We therefore used the dipole parameters obtained from fitting single equivalent dipoles to the individual component loadings as start values for the multiple dipole model.

The multiple dipole model therefore can be formulated as follows:

$$\sum_{i=1}^{ntim} \sum_{j=1}^{nmea} \{ y_{ij} - \sum_{k=1}^{nsrc} a_{ki} * f_j (x_k, y_k, z_k, xc_k, yc_k, zc_k) \}^2 \longrightarrow min$$

with y_{ij} = electric potential or magnetic field at measurement site j at time point i, a_{ki} = activity of source k at time point i, $f_j (x_k, y_k, z_k, xc_k, yc_k, zc_k)$ = electric potential or magnetic field, which is generated by a unit dipole with location (x_k, y_k, z_k) and components $\{(xc_k, yc_k, zc_k), xc_k^2 + yc_k^2 + zc_k^2 = 1\}$ at measurement site j;

$$i = 1, ..., ntim: \text{number of time points}$$
$$j = 1, ..., nmea: \text{number of measurement sites}$$
$$k = 1, ..., nsrc: \text{number of dipole sources}$$

3.7. Practical Outline of Spatiotemporal Modeling

The practical steps involved in spatiotemporal modeling can be summarized as follows. First, PCA was applied to the data. The number of significant principal components or brain sources underlying the data was estimated by the 'eigenvalue greater-one criterion' and by the 'variance criterion' (Section 3.4.2). For these significant principal components, the factor loadings were displayed as spatial patterns, i.e. isocontour plots, and the factor scores as time activities. The spatiotemporal patterns of the principal components were calculated by multiplying the factor loadings with the factor scores. The spatiotemporal pattern of the individual principal components was summed up to the combined spatiotemporal pattern of all significant principal components. In the next step, single equivalent dipole fits were performed on the factor loadings representing the spatial potential or field distributions of the hypothetical brain sources. Finally, the multiple dipole model was applied. The dipole parameters obtained from the single dipole fits to the principal components factor loadings served as start values for the non-linear minimization procedure. Furthermore, the hypothesis concerning the number of underlying brain sources was retested as the multiple dipole model was applied separately with different numbers of sources. The contribution of a given dipole was considered significant if it increased the amount of variance explained by the model more than run-to-run noise.

3.8. Spatiotemporal Modeling as Outlined on Two Typical Examples

In the following, we will present two typical examples to outline the procedure of spatiotemporal modeling. This should provide a better understanding of the method. As a first example, we present spatiotemporal modeling as applied to somatosensory evoked magnetic fields which is especially relevant to this study. The second example deals with the application of spatiotemporal modeling to the human epileptic spike complex on MEG and was chosen to demonstrate that this approach may also be useful to study pathophysiology of human disease.

3.8.1. Modeling of Somatosensory Evoked Magnetic Fields

Fig. 3.4A shows the MEG recording matrix and the somatosensory evoked magnetic fields (SEFs) recorded from 45 measurement sites (9 different, overlapping positions of a 7-channel magnetometer yielding a total of 63 and 45 distinct measurement sites as indicated). A detailed description of the MEG recording strategy is given in Section 4.1.3. PCA indicated two significant principal components underlying the SEFs. The results of PCA are displayed in Figs. 3.4B – 3.4E. The component loadings represent the spatial field patterns generated by the principal components and are shown in form of isofield maps (Fig. 3.4B). The component scores

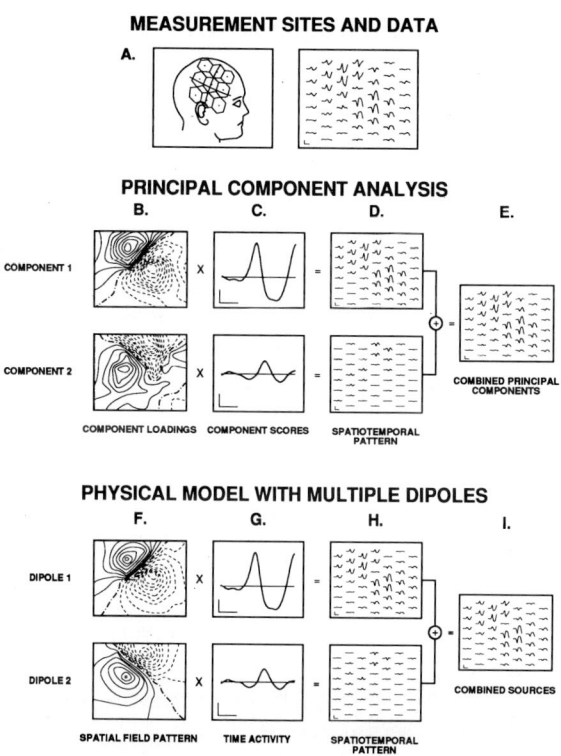

Fig. 3.4. Spatiotemporal modeling of somatosensory evoked magnetic fields (SEFs).
A. MEG recording matrix and SEFs recorded from 45 measurement sites. **B-E.** Results
of PCA, showing component loadings (B), component scores (C), as well as
spatiotemporal patterns of the individual (D) and the combined principal
components (E). **F-I.** Results of multiple dipole modeling, showing spatial field
patterns (F), time activities (G), as well as spatiotemporal patterns of the individual
(H) and the combined dipole sources (I). Both PCA (E) and multiple dipole
modeling (I) could reproduce the data (A) very closely. Calibration: horizontal = 20
msec, time scale begins at stimulus onset; vertical = 100 fT (A, D, E, H, and I);
horizontal = 10 msec, time scale begins at stimulus onset; vertical = 50 fT (C and G);
isofield maps: 10% isocontour lines, each map scaled to 1 fT

correspond to the time activities of the principal components (Fig. 3.4C).
The spatiotemporal pattern of a given principal component can be
calculated by multiplying the component loadings with the component
scores (Fig. 3.4D). The first principal component explained 77% and the
second 17% of the data variance. The spatiotemporal patterns of the
individual principal components (Fig. 3.4D) sum up to the combined
spatiotemporal pattern (Fig. 3.4E). This pattern could reproduce the actual
data very closely accounting for 94% of the data variance (compare Figs.
3.4A and 3.4E).

In the next step, we applied single dipole fits on the component
loadings. The obtained dipole parameters served as start values for the
multiple dipole model. Multiple dipole modeling was performed using two

and three dipoles. Adding a third dipole did not significantly increase the variance accounted for by the model. Thus, in agreement with the results of PCA, a two dipole model was used. The results of multiple dipole modeling are shown in Figs. 3.4F – 3.4I. Similar to PCA, the spatiotemporal field patterns of the individual dipoles (Fig. 3.4H) can be obtained by multiplying the spatial field patterns (Fig. 3.4F) with the time activities (Fig. 3.4G). The first dipole accounted for 70%, the second for 21% of the data variance. The combined spatiotemporal pattern (Fig. 3.4I) was obtained by summing up the field contributions of the individual dipoles (Fig. 3.4H) and could reproduce the data very closely accounting for 89% of the data variance (compare Figs. 3.4A and 3.4I). As the sources were not orthogonal, the variance contributions of the individual dipoles do not simply sum up to the variance explained by the model like the principal components.

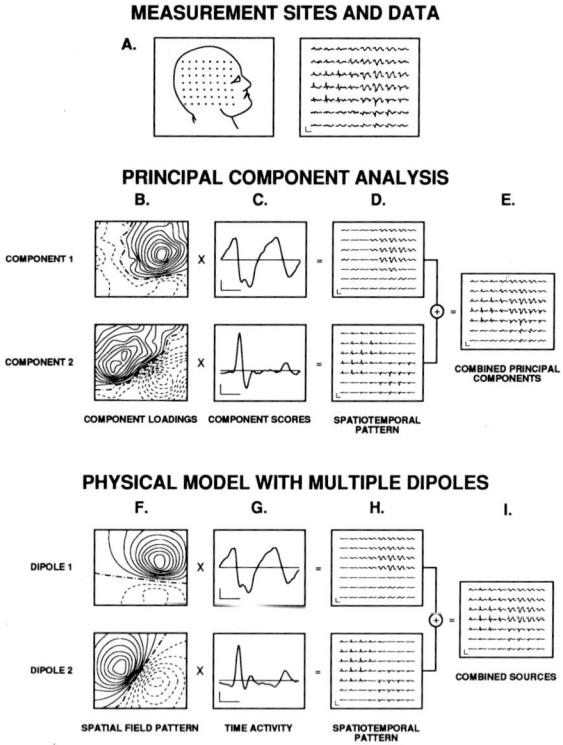

Fig. 3.5. Spatiotemporal modeling of the human epileptic spike complex. **A.** MEG recording matrix and epileptic spike complex recorded from 56 measurement sites. **B-E.** Results of PCA, showing component loadings (B), component scores (C), as well as spatiotemporal patterns of the individual (D) and the combined principal components (E). **F-I.** Results of multiple dipole modeling, showing spatial field patterns (F), time activities (G), as well as spatiotemporal patterns of the individual (H) and the combined dipole sources (I). Both PCA (E) and multiple dipole modeling (I) could reproduce the data (A) very closely. Calibration: horizontal = 0.5 sec; vertical = 500 fT (A, D, E, H, and I); horizontal = 0.25 sec; vertical = 250 fT (C and G); isofield maps: 10% isocontour lines, each map scaled to 1 fT

Although the results of PCA and multiple source modeling looked similar, there were some differences especially between component 2 and dipole 2 (Figs. 3.4B and 3.4F). This can be explained by the fact that PCA is a purely statistical technique bound only to mathematical constraints (e.g. the orthogonality condition) without any physiological restrictions. Multiple source modeling, on the contrary, is subject to a dipolar source configuration, but the sources do not need to be orthogonal to each other.

3.8.2. Modeling of the Human Epileptic Spike Complex

MEG measurements were made with a single channel magnetometer from 56 distinct measurement points (Fig. 3.5A). At least 15-20 MEG spikes were averaged for each measurement location using the scalp-EEG spikes as triggers according to the spike averaging technique introduced by Barth et al. [27, 28, 29]. The epileptic spike complex is shown in Fig. 3.5A. The modeling procedure consisted of the same basic steps as already discussed in the previous example on somatosensory evoked magnetic fields, i.e. the application of PCA, of single dipole modeling on the component loadings, and finally of multiple dipole modeling. The results of PCA – spatial component loadings, component scores, as well as individual and combined spatiotemporal patterns – are shown in Figs. 3.5B – 3.5E. Multiple source modeling yielded dipoles in the anterior and midtemporal region 26 mm and 38 mm beneath the scalp. The time activities of the sources in this particular patient indicated that the two components of the spike complex, i.e. the spike and the slow wave, were generated by two spatially distinct neuronal populations. Both PCA (Fig. 3.5E) and multiple source modeling (Fig. 3.5I) could reproduce the actual data set (Fig. 3.5A) very closely and explained over 85% of the data variance. Analogous to the previous example, results of PCA and multiple dipole modeling were similar, but not identical.

3.9. Limitations of the Procedure

3.9.1. Limitations of Principal Component Analysis

PCA is a statistical technique with no relation to the underlying brain physiology. The basic modeling assumptions of successively accounting for a maximum of the system variance and the orthogonality condition are not physiologically meaningful. This can also be seen by comparing the results of PCA to those of multiple source modeling in the previous two examples (Figs. 3.4 and 3.5). Thus, it is questionable whether the individual principal components represent electric potential or magnetic field contributions generated by distinct brain regions or merely a mixture of these sources that cannot be separated. Thus, the physiological meaning of the results of principal component analysis has been under extensive debate [40, 140, 247, 308, 363, 376, 377].

In order to improve the interpretability of the results of principal component analysis and to better represent the actual brain sources by the individual principal components, the coordinate axes of the common factor space (principal components) can be rotated without any effect on the adequacy of the solution [167], see also Section 3.4.2. Thus, various methods of rotation have been applied to enhance biological interpretability of the results of PCA. The most common method of rotation applied in neurophysiological research has been the 'varimax rotation' [40, 140, 247, 308, 363, 376, 377]. Other techniques include the 'oblimin rotation' [112, 167] and the 'frequency domain rotation' introduced to neurophysiological research by Baumgartner et al. [40]. Advantages and shortcomings of these different techniques have been investigated in a computer simulation study [40] and go beyond the scope of this chapter. In short, this study showed that each of these procedures can be useful for the separation of brain activity generated by different neuronal populations, but also can fail in special situations. Generally, it is impossible to decide a-priori which method of rotation will perform best. However, a comparison of the results of the different methods of rotation is useful to enhance biological reliability and interpretability of the results of PCA.

In conclusion, PCA should be viewed more as a tool to determine the number of sources underlying the measured brain activity and to form an initial working hypothesis, rather than as a method of making definite conclusions concerning brain physiology.

3.9.2. Limitations of Multiple Dipole Modeling

Limitations concerning multiple source modeling will be discussed in conjunction with the results of spatiotemporal modeling of somatosensory evoked responses in Section 4.5.3.

4. Functional Anatomy of Human Somatosensory Cortex

4.1. Methods

4.1.1. Cortical Stimulations – Median Nerve Somatosensory Evoked Potentials on Electrocorticography

4.1.1.1. Patients

We studied 6 epilepsy patients who were evaluated for epilepsy surgery. All patients had partial seizures defined by EEG recordings and behavior during seizures. The seizures were medically intractable and severely interfering with the patients' quality of lives. Non-invasive studies (scalp and sphenoidal-EEG telemetry, neuropsychological evaluation, functional imaging studies (PET), and structural imaging studies (CT, MRI)) did not allow a definitive localization of the seizure focus. Thus, the patients were evaluated with chronically indwelling subdural grid electrodes for definitive localization of the seizure focus. Table 4.1.1 (page 56) shows the patients' clinical data and seizure types. All patients had given written informed consent to undergo invasive presurgical evaluation and to participate in the SEP studies.

4.1.1.2. Cortical Stimulations

We performed cortical stimulations in all patients from chronically indwelling subdural grid electrodes. Details of this procedure are outlined in Section 2.4. Subdural grids were implanted to delineate essential cortex, i.e. motor, sensory, and speech cortex, and to record interictal as well as ictal epileptiform discharges to exactly localize the seizure focus.

In all patients, large grids containing 64 electrodes were placed over the lateral cortical surface, and smaller grids (containing 10-20 electrodes) and strips (containing 4-8 electrodes) under the temporal and frontal lobe as well as in the interhemispheric fissure. The electrode grids and strips consisted of platinum iridium discs with a diameter of 6 mm and 10 mm spacing center-to-center embedded in a silastic sheet (PMT Corporation, Minneapolis, MN).

Grids were implanted together with an intracranial pressure (ICP) monitor usually on a Thursday or Friday. After the implantation, patients were transferred to the neurosurgical intensive care unit (ICU) and monitored overnight. On the next morning, patients were transferred to the epilepsy monitoring unit (EMU). Cortical stimulations were not begun till Monday to give the brain time to 'adapt' to the grids implanted [218].

Subsequently, cortical stimulations and SEP studies were performed during the first week of presurgical evaluation. The patients were still on high levels of antiepileptic drugs thus reducing afterdischarge threshold. Cortical stimulations were performed for an average of 8 hours a day and were completed within 5 days. To reduce fatigue, rest periods were interspersed with the stimulations. Each electrode site with positive or negative responses (either motor, sensory, or speech) was stimulated twice to test for reproducibility of the results. A response was classified as primary motor when positive motor responses were obtained at low stimulation intensities (around 2-4 mA) with 50 Hz trains, and furthermore single pulse stimulation yielded a one-to-one response of muscle twitches. Cortical stimulations were considered as obtained from primary somatosensory cortex when sensory experiences were reported by the patient before any other effect could be observed by the investigator, i.e. a muscle twitch. Furthermore, these responses were required to occur at low stimulation intensities usually around 3-5 mA. After completion of the cortical stimulations, antiepileptic drugs were reduced to provoke seizures for definitive localization of the seizure focus.

4.1.1.3. Somatosensory Evoked Potentials on Electrocorticography

We measured evoked potentials on electrocorticography (ECoG) during median nerve stimulation from these chronically indwelling subdural grid electrodes in all 6 patients. ECoG was measured during the first 60 msec after shock stimulation of the contralateral median nerve at the wrist. The stimulus was delivered by a Grass S88 stimulator (Grass Instrument Company, Quincy, MA) and consisted of monophasic, constant current 0.3 msec pulses which produced a visible thumb twitch. ECoG was recorded simultaneously from a rectangular array of 48 chronically indwelling subdural grid electrodes with a diameter of 6 mm and 10 mm spacing center-to-center (PMT Corporation, Minneapolis, MN). ECoG was referenced to a scalp-EEG needle electrode at central vertex. Data were amplified (x 10,000) and filtered (bandpass 1 to 1,000 Hz) using Grass 12A5 amplifiers (Grass Instruments Company, Quincy, MA), digitized at a sampling rate of 4096 Hz (12 bits), and stored digitally for off-line data analysis. Two runs of 250 trials each were superimposed to assess reproducibility.

4.1.1.4. Correlation of Neuroelectric and Anatomical Data

In order to get an exact correlation of electrophysiological and anatomical data, we mapped the source localizations unto composite diagrams of cortex obtained from intraoperative photographs, skull x-rays, magnetic resonance images, and cortical stimulations. The precise location of central sulcus was determined by comparing intraoperative photographs with the electrodes in place with the results of cortical stimulations and then superimposing the intraoperative photographs with the skull x-rays. We confirmed that the electrodes did not move by using repeated intraoperative photographs and extraoperative skull x-rays. We constructed a two-dimensional cortical map for each patient. On these cortical maps, source localizations were depicted as surface projections, i.e. perpendicular projection of the source from the center of the sphere unto the cortical surface. All patients had a detailed neurological examination before and after resective surgery. During surgery somatosensory cortex as outlined by cortical stimulations was spared.

4.1.1.5. Data Analysis

Reproducibility of the data was tested by calculating run-to-run variability as the sum of squared differences between run #2 and run #1 divided by the sum of squares of run #1.

In order to investigate the spatiotemporal structure of the SEPs over the entire time domain of the evoked response we applied spatiotemporal modeling. The theoretical and technical details of spatiotemporal modeling are described in Section 3.

The forward solution was obtained using a homogeneous spherical model [83, 84, 337, 338, 372]. We assumed that inhomogenities would have little effect on modeling since recordings were made directly from the surface of the brain. Furthermore, the surface of the brain is similar to a sphere in the parieto-frontal region. The radius of the sphere was determined from MRI scans, the conductivity of the brain was assumed as $(0.33 \ (\Omega m)^{-1})$ according to values reported in the literature [83, 136, 337]. The shape of the central sulcus was not taken into account in the model as the effect of fissures has been shown to produce only small distortions of the electric fields outside a spherical volume conductor [80]. Thus, each dipole was given uniquely by 6 parameters (3 location parameters and 2 orientation parameters remaining constant over time and determining the spatial potential pattern, and one strength parameter representing the time varying activity). For further details see also the Section 2.2.

The basic modeling assumption was that the sources were fixed in space and their strength was varied over time. The goal of the modeling procedure was thus to determine which combination of multiple spatial dipole patterns, weighted by the respective time activities, yielded a combined spatiotemporal pattern that most closely reproduced the data. For a given set of dipole location and orientation parameters which determined the spatial potential patterns the time activities were computed using linear regression analysis [118]. The dipole location and orientation parameters were varied iteratively using the simplex algorithm [295] until the optimum combination of source location, orientation, and time activity parameters was obtained. The start values for the non-linear minimization procedure were obtained as described in Sections 3.5 and 3.7.

Variance accounted for was calculated as the sum of squared differences between the data and the model divided by the sum of squares of the data across all recording positions and all time points.

In order to determine the number of sources incorporated in the multiple dipole model, we applied two independent criteria as already described in Sections 3.5 and 3.7. First, we applied principal component analysis to determine the number of independent underlying covarying parts of the system variance. Second, we used multiple dipole models with two and three dipoles. If adding an additional dipole did not significantly increase the amount of variance explained – i.e. was smaller than noise (variability between successive runs) – then adding this source was considered insignificant.

4.1.2. Median Nerve Somatosensory Evoked Potentials on Scalp-EEG

4.1.2.1. Subjects and Procedures

We measured somatosensory evoked potentials (SEPs) on scalp-EEG during the first 60 msec after shock stimulation of the median nerve at the wrist in 4 normal healthy volunteers and one epilepsy patient who was evaluated for a definitive localization of the seizure focus (total of 5 subjects). All subjects had given written informed consent to participate in the study.

The stimulus was delivered by a Grass S88 stimulator (Grass Instrument Company, Quincy, MA) and consisted of monophasic, constant current 0.3 msec pulses which produced a visible thumb twitch. Thus, stimulus parameters were identical to those used for median nerve SEPs recorded on ECoG (see Section 4.1.1.3).

Scalp-EEG was recorded simultaneously from a rectangular array of 32 (2 subjects) or 48 (3 subjects) gold disc electrodes (Grass Instrument Company, Quincy, MA). EEG was referenced to linked ears providing an uninvolved reference. The 32 electrodes were arranged in a 6 times 6 matrix centered at the C3 or C4 position of the International 10-20-System (2 electrodes each at the posterior superior and anterior inferior corners of this grid were omitted). The 48 electrodes were arranged in a 7 times 7 grid centered at the C3 or C4 position of the International 10-20-System (one electrode at the anterior inferior corner of this grid was omitted). The electrodes were spaced 3 cm center-to-center. Data were amplified (x 100,000) and filtered (bandpass 1 to 1,000 Hz) using Grass 12A5 amplifiers (Grass Instruments Company, Quincy, MA), digitized at a sampling rate of 4000 Hz (12 bits), and stored digitally for off-line data analysis. Two runs of 500 trials each were superimposed to assess reproducibility.

4.1.2.2. Correlation of Neuroelectric and Anatomical Data

In order to determine the exact location of the electrodes in relation to anatomical landmarks, skull x-rays with the electrodes in place were taken in two planes (anterior-posterior and lateral views). Additionally, for each subject, a head coordinate system was defined by 3 anatomical landmarks (left and right preauricular points, and nasion) using a probe positioning system (Ptolhemus Navigation Sciences, McDonnel Douglas, Colchester, Vermont, USA). The head coordinate system was defined by the x-axis passing through the left and right preauricular points, and the y-axis bisecting this line and passing through the nasion. The z-axis was chosen perpendicular to the x-y-plane and approximately passed through the vertex, such that the coronal plane (x-z-plane) was approximately parallel and slightly posterior to central sulcus. Using the probe positioning system, the three-dimensional location of the electrodes in this head coordinate system and thus in relation to anatomical landmarks on the subject's head could be determined.

4.1.2.3. Data Analysis

Similar to the ECoG study (Section 4.1.1.5), reproducibility of the data was determined by calculating run-to-run variability. To investigate the spatiotemporal structure of the SEPs over the entire time domain we applied spatiotemporal modeling using multiple dipoles. The details of spatiotemporal modeling are described in Section 3.

We used a 4-shell spherical model with different conductivities for the brain tissue (0.33 $(\Omega m)^{-1}$), the cerebrospinal fluid (0.0042 $(\Omega m)^{-1}$), the skull (1.0 $(\Omega m)^{-1}$), and the scalp (0.33 $(\Omega m)^{-1}$) to obtain the forward solution [84, 338]. The conductivity values were chosen in accordance with reports in the literature [15, 83, 136, 337]. The outermost radius for the scalp was obtained from the best-fitting sphere within the electrode positions determined by the probe positioning system (PPI). The next two radii corresponding to the outer and inner surface of the skull were achieved from the skull x-rays. The radius of the brain finally was found empirically by comparing the outer radii with values in the literature and setting the brain radius to an appropriate proportion. We chose a spherical model as the surface of the skull is similar to a sphere in the parieto-frontal region. Thus, each dipole was uniquely given by 6 parameters (3 location and 2 orientation parameters remaining constant over time and determining the spatial potential pattern, and one strength parameter representing the time varying activity). For further details see also the Section 2.2.

In complete analogy to the data analysis of median nerve SEPs on ECoG presented in Section 4.1.1.5, the goal of the modeling procedure was to find those dipoles with fixed anatomical location resp. orientation and time varying activity which most closely reproduced the spatiotemporal structure of the data. Start values for the non-linear minimization problem were obtained according to the procedures described in Sections 3.5 and 3.7. The number of sources used in the multiple dipole model was determined by principal component analysis and by testing multiple dipole models with different numbers of dipoles and applying the variance criterion (Sections 3.5 and 3.7).

4.1.3. Median Nerve Somatosensory Evoked Fields on Magnetoencephalography

4.1.3.1. Subjects and Procedures

We measured somatosensory evoked magnetic fields (SEFs) during the first 60 msec after shock stimulation of the median nerve at the wrist in 9 healthy subjects. All subjects had given written informed consent to participate in the experiment. The stimulus was delivered by a Grass S88 stimulator (Grass Instrument Company, Quincy, MA) and consisted of monophasic, constant current 0.3 msec pulses at 3.1 Hz to produce a visible thumb twitch and thus was identical to the stimulus parameters used for the ECoG and scalp-EEG recordings. Neuromagnetic recordings were performed with a seven channel magnetometer with dc-SQUIDs and coplanar coils (second derivative gradiometer; coil diameter 15 mm; coil baseline 40 mm; Biomagnetic Technologies Inc., San Diego, CA) in a large electromagnetically shielded room (Vacuumschmelze GMBH, Hanau, FRG). Magnetometer coils were located in a plane in a hexagonal array with 27 mm distance center-to-center in a dewar with a flat bottom. MEG system noise was 2.5 picotesla (16 femtotesla/√Hz). The magnetometer was placed at 9 overlapping positions yielding a total of 63 measurement points and 45 distinct measurement points. The measurement matrix was centered at the C3 or C4 position of the International 10-20-System. Data were filtered with a bandpass of 1 to 1,000 Hz, digitized at a sampling rate of 4096 Hz (12 bits), and stored digitally for off-line data analysis. Two runs of 500 trials each were superimposed to assess reproducibility.

The subjects were lying comfortably on a wooden non-magnetic bed. The subject's head and upper body were fixated with vacuum casts to prevent movement during the experiment. Furthermore, the stability of the position of the subject's head in relation to the magnetometer was determined by assessing superimposition of a plastic sheet attached to the subjects's head and the bottom of the dewar before and after each experimental session (see also Section 4.1.3.2).

4.1.3.2. Correlation of Neuromagnetic and Anatomical Data

In order to determine the relation of neuromagnetic and anatomical data, the exact magnetometer position in relation to the subject's head had to be determined. This was achieved by superimposing the magnetometer with a plastic sheet sized identical to the bottom of the dewar which was attached tangentially to the subject's head at the center of each probe position. These points were marked and skull x-rays were taken subsequently in two planes (anterior-posterior and lateral views) with the marks in place.

This procedure was performed because no probe positioning system was available when we performed this study. Thus, the correlation of neuromagnetic and anatomical data is somewhat different compared to that of the study on somatotopy on MEG presented in Section 5.1.3.2.

4.1.3.3. Data Analysis

Data analysis was similar to that already described for median nerve SEPs recorded on ECoG (Section 4.1.1.5) and scalp-EEG (Section 4.1.2.3). Reproducibility of the data was determined by calculating run-to-run variability. We analyzed the spatiotemporal structure of the SEFs by multiple dipole modeling. For details of spatiotemporal modeling see Section 3.

We used a homogeneous spherical model to obtain the forward solution. A homogeneous sphere was chosen because magnetic fields are not influenced by concentric changes in conductivity [88, 314, 369, 370]. The radius of the sphere was determined from the position of the markers on the skull x-rays. The assumption of a spherical volume conductor does not introduce significant errors in source localization in the parieto-frontal region for neuromagnetic measurements [157]. Thus, each dipole was uniquely given by 5 parameters: 3 location parameters representing its intracerebral location and one orientation parameter representing its orientation in a plane tangential to the surface of the sphere. As already mentioned in Section 2.2, in contrast to electric measurements where 6 dipole parameters are necessary, in MEG 5 dipole parameters suffice as the radial component of a dipole generates no measurable magnetic field outside a spherical volume conductor [150]. These parameters remained constant over time and determined the spatial magnetic field pattern. Additionally, one strength parameter represented the time varying activity of each dipole.

Analogous to the procedure performed on median nerve SEPs recorded on ECoG (Section 4.1.1.5) and scalp-EEG (Section 4.1.2.3), the goal of the modeling procedure was to find that set of multiple dipoles with fixed location and orientation and time varying activity which most closely reproduced the data. Start values for the non-linear minimization procedure were obtained as described in Sections 3.5 and 3.7. The number of dipoles was determined by principal component analysis and by performing multiple dipole modeling separately with two and three dipoles and applying the variance criterion (see Sections 3.5 and 3.7).

4.2. Results

4.2.1. Cortical Stimulations – Median Nerve Somatosensory Evoked Potentials on Electrocorticography

4.2.1.1. Cortical Stimulations

Figs. 4.2.1.1 and 4.2.1.3 show skull x-rays with the subdural grid electrodes in place, and the results of cortical stimulations for Patients #1 and #2. Central sulcus is shown in relation to the subdural electrodes. Whereas electrodes with positive motor responses are depicted as black, electrodes with positive sensory responses are depicted as gray, and electrodes with mixed motor/sensory responses as half black and gray.

Electrodes with negative stimulation effects (e.g. negative motor responses), with speech interference, and without stimulation effect were not differentiated and are shown as white.

Stimulation of electrodes anterior to central fissure (corresponding to Brodman's area 4) yielded positive motor responses. Thus, stimulation of these electrodes with stimulus trains of 50 Hz resulted in sustained positive motor responses at very low stimulation intensities, usually around 2-4 mA. Furthermore, single pulse stimulation elicited one-to-one motor responses consisting of muscle twitches. Posterior to central fissure primary sensory responses could be elicited from electrodes overlying Brodman's areas 1 and 2. Stimulation intensities still were low around 3-5 mA, but slightly higher compared to those for primary motor responses. Additionally, from some electrodes mixed motor/sensory responses could be obtained. A response was considered mixed motor/sensory when both motor and sensory responses could be elicited at the same low stimulation intensities, and despite repeated stimulations it could not be decided whether to classify the response as motor or sensory.

It should be stressed that the separation of motor and sensory responses by central sulcus was not absolute. From some electrodes anterior to central sulcus sensory responses could be obtained, whereas stimulation of some electrodes posterior to central sulcus yielded motor responses. Furthermore, both motor and sensory responses could be elicited not only from electrodes immediately adjacent to central fissure, but also from a broad band extending several centimeters in the rostrocaudal direction both anterior and posterior to central fissure. Responses were strictly contralateral. The quality of sensation induced by cortical stimulations was usually described as tingling or numbness.

4.2.1.2. Somatosensory Evoked Potentials – Data

Median nerve SEPs were reproducible within patients showing a run-to-run variability of 5-8%. Between patients there was variability in waveforms and peak latencies which could be attributed to age differences and differences in lesion sites (compare Figs. 4.2.1.2D and 4.2.1.4D).

Median nerve SEPs showed prominent peaks at latencies of about 20 and 30 msec. Anterior to central fissure positive-negative waveforms were recordable, posterior to central fissure negative-positive waveforms were recordable. Corresponding to their polarity and latency, these waveforms have been termed P20-N30 components (anterior to central fissure) and N20-P30 components (posterior to central fissure) [7, 8, 117]. For sake of simplicity, we will refer to these components as N20 and P30. Both N20 and P30, showed a clear phase reversal across central sulcus. These waveforms can be identified very clearly for Patient #1 (see Fig. 4.2.1.2D: column 5, rows 2-4; column 6, rows 2 and 3; column 7, row 4; columns numbered from left to right; rows numbered from top to bottom) and less clearly for Patient #2 (see Fig. 4.2.1.4D: column 6, rows 2 and 3; column 8, row 3).

In addition, positive peaks were recorded at about 25 msec and negative peaks at about 35 msec. According to their polarity and latency, these components have been termed P25 and N35 components [7, 8, 117]. These components did not show a phase reversal across central sulcus, but instead showed single electropositivities and electronegativities, respectively. These components were very prominent in Patient #2 (see Fig. 4.2.1.4D: column 7, row 2; column 8, row 2), but could be hardly identified in Patient #1 (see Fig. 4.2.1.2D: column 7, row 3).

4.2.1.3. Number of Sources

First, we applied principal component analysis (PCA) to determine the number of underlying covarying processes. Two principal components accounted for 88.9% (range: 81.1-95.9%) and three principal components for 91.9% (range: 83.4-98.2%) of the data variance. Thus, PCA indicated that there were two significant processes underlying the system variance.

In a second step, we applied multiple dipole models with two and three dipoles. Two sources accounted for an average of 84.6% of the data variance (range: 76.3-89.5%). Adding a third source explained less variance than run-to-run variability.

Therefore, we used a model with two dipoles for further analysis.

4.2.1.4. Results of Spatiotemporal Modeling

The results of spatiotemporal modeling were qualitatively similar in all patients. The peak latencies averaged 20 msec (range: 18-22 msec) as well as 30 msec (range: 28-32 msec) for the first source, and 24 msec (range: 22-27 msec) as well as 34 msec (range: 32-38 msec) for the second source, which corresponded closely to the N20-P30 and P25-N35 peaks, respectively. Therefore, we will refer to the first source as N20-P30 source and to the second source as P25-N35 source in the following. Both sources were located in postcentral gyrus near electrodes where cortical stimulations produced hand sensations. The distance between the surface projection of the N20-P30 source (radial projection of the three-dimensional intracerebral source location from the center of the sphere unto the cortical surface) and the central sulcus averaged 6 mm (range: 3-10 mm). The distance between the surface projection of the P25-N35 source and central sulcus averaged 7 mm (range: 4-9 mm), the distance between both sources was 10 mm (range: 8-11 mm). Concerning the medio-lateral direction, the N20-P30 source was located lateral to the P25-N35 source.

The mean depth of both sources in the patient population was about 8 mm. Thus, there were no significant differences concerning depth between the N20-P30 source and the P25-N35 source (mean depth of the N20-P30 source: 8 mm below the cortical surface, range: 6-10 mm; mean depth of the P25-N35 source: 8 mm below the cortical surface, range: 4-10 mm). Concerning orientation, the N20-P30 source was oriented in the anterior posterior direction and its spatial potential pattern showed a phase reversal across central sulcus in all patients. The P25-N35 source, on the contrary,

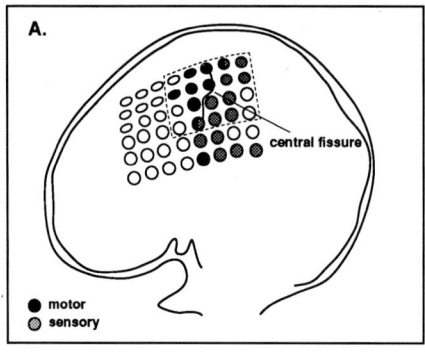

SKULL X-RAY AND ELECTRODE GRID

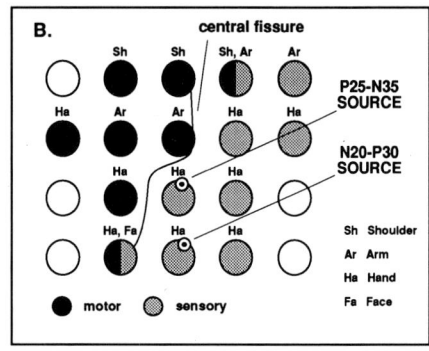

SOURCE LOCATIONS / CORTICAL STIMULATIONS

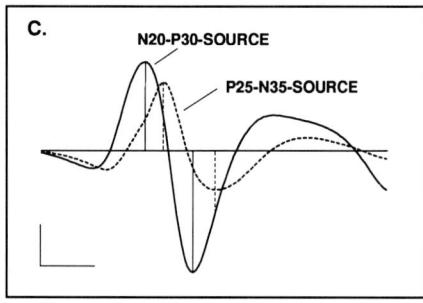

TIME PATTERN

Fig. 4.2.1.1. Patient #1. Skull x-ray with grid electrodes, locations and time activities of the sources, and results of cortical stimulations. **A.** Skull x-ray with electrode grid. Region inside dotted rectangle is enlarged for clarity to right in (B). **B.** Enlargement shows source locations and results of cortical stimulations. Motor and sensory responses elicited by cortical stimulations as well as central fissure are shown. Surface projections of source locations are displayed as circled dots. Both sources are in hand sensory cortex near a bend in central fissure. **C.** Superimposed time patterns of the N20-P30 source and the P25-N35 source. Vertical lines indicate peak latencies. Polarities: positive (+) up, negative (−) down. Calibration: horizontal = 10 msec, time scale begins at stimulus onset; vertical = 10 μV

showed no consistent orientation across patients. In some patients, this source showed a primarily radial pattern (cf. Patient #2), whereas in others it showed significant tangential components (cf. Patient #1). In all patients, the potentials generated by the two sources overlapped substantially both in space and time.

4.2.1.5. Results of Selected Patients

In the following, we present the results of selected patients. Figs. 4.2.1.1 and 4.2.1.2 show the data and results of Patient #1. Fig. 4.2.1.1 summarizes the functional anatomy and different time courses of the two sources. Both sources were located in postcentral gyrus about 8 mm beneath the cortical surface and next to electrodes where cortical stimulations produced hand sensations. In Fig. 4.2.1.1B the sources are depicted as surface projections, i.e. as radial projections of the three-dimensional intracerebral source

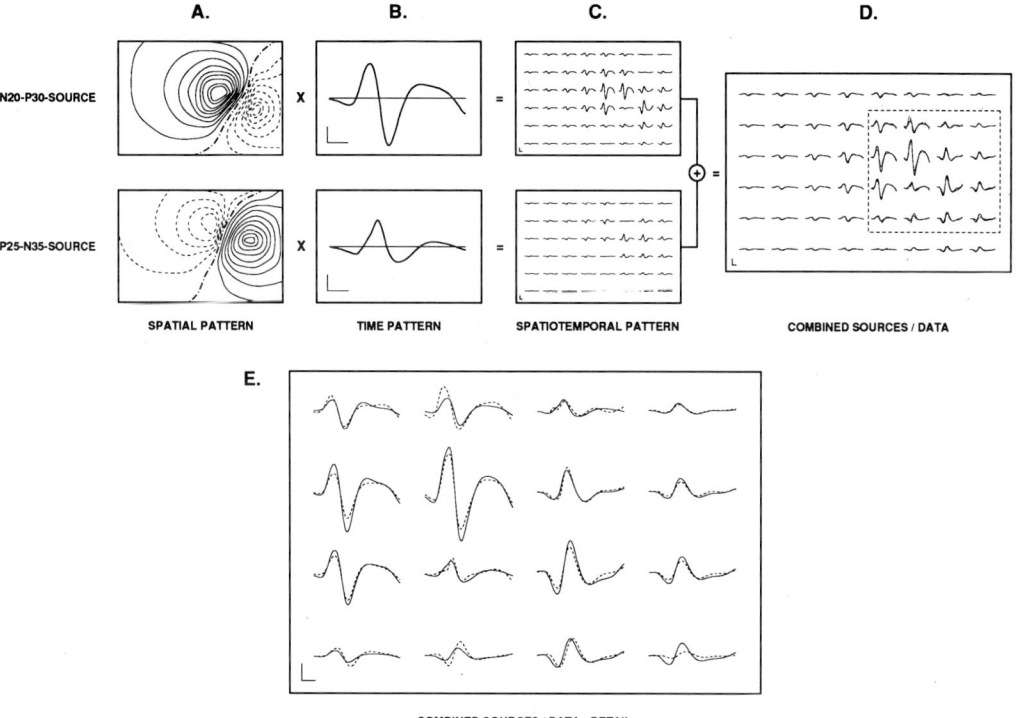

Fig. 4.2.1.2. Patient #1. Spatiotemporal structure of the somatosensory evoked response on ECoG. **A.** Spatial patterns of the individual sources shown as isopotential maps (10% isocontour lines; maximum of each map = 1 µV; polarities: positive (+) solid, negative (–) dashed). **B.** Time patterns of the individual sources. **C.** Spatiotemporal patterns of the individual sources, which is (A) times (B). **D.** Superposition of the combined spatiotemporal pattern of the two sources (solid) and the original data (dashed). **E.** Enlargement of region inside dotted rectangle in (D), showing superposition of combined spatiotemporal pattern (solid) and the original data (dashed). Calibration: horizontal = 10 msec, time scale begins at stimulus onset; vertical = 10 µV

location from the center of the sphere unto the cortical surface. The distance from central sulcus was 8 mm for the N20-P30 source and 4 mm for the P25-N35 source. The N20-P30 source was located lateral inferior to the P25-N35 source, and the distance between both sources was 10 mm. The N20-P30 source had peak latencies at 19.3 and 28.1 msec, and the P25-N35 source at 22.5 and 32.4 msec (Fig. 4.2.1.1C). Especially at latencies of 25 and 35 msec, there was considerable overlap of the time activities of the two sources (Fig. 4.2.1.1C).

Fig. 4.2.1.2 shows the spatiotemporal structure of the evoked response in this patient. Concerning the spatial patterns, the N20-P30 source showed a tangential potential pattern with a phase reversal across central fissure. The P25-N35 source was slightly tilted towards radial compared to the N20-P30 source, but also showed a primarily tangential potential pattern. The time

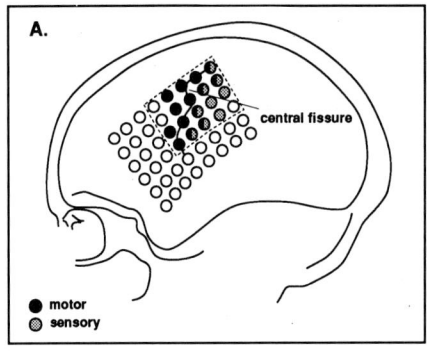

SKULL X-RAY AND ELECTRODE GRID

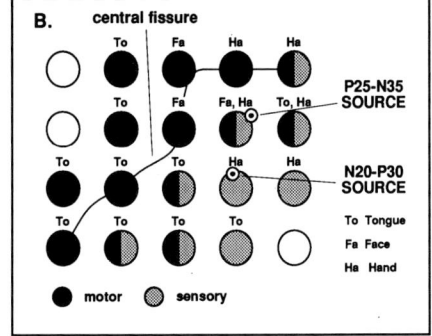

SOURCE LOCATIONS / CORTICAL STIMULATIONS

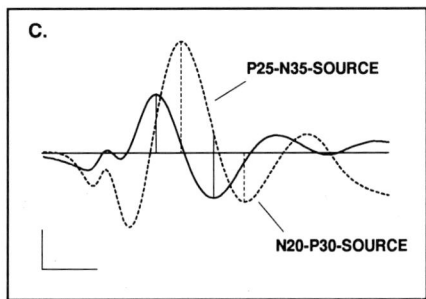

TIME PATTERN

Fig. 4.2.1.3. Patient #2. Skull x-ray with grid electrodes, locations and time activities of the sources, and results of cortical stimulations. **A.** Skull x-ray with electrode grid. Region inside dotted rectangle is enlarged for clarity to right in (B). **B.** Enlargement shows source locations and results of cortical stimulations. Motor and sensory responses elicited by cortical stimulations as well as central fissure are shown. Surface projections of source locations are displayed as circled dots. **C.** Superimposed time patterns of the N20-P30 source and the P25-N35 source. Vertical lines indicate peak latencies. In comparison to Patient #1 (Fig. 4.2.1.1C), in this patient the two sources showed more overlap and also additional peaks at about 15 msec. Polarities: positive (+) up, negative (–) down. Calibration: horizontal = 10 msec, time scale begins at stimulus onset; vertical = 10 µV

activities are shown in Fig. 4.2.1.2B. The spatiotemporal patterns of the individual sources (Fig. 4.2.1.2C) were calculated by multiplying the spatial patterns (Fig. 4.2.1.2A) with the time patterns (Fig. 4.2.1.2B). It should be noted that in this patient the contribution of the N20-P30 source to the evoked response was much larger than that of the P25-N35 source (compare Fig. 4.2.1.2C top and bottom).

The spatiotemporal pattern of the combined sources (Fig. 4.2.1.2D) was obtained by summing up the individual spatiotemporal patterns (Fig. 4.2.1.2C) according to the law of superposition. The spatiotemporal patterns of the individual sources showed considerable overlap and could not readily be identified from visual inspection of the raw data. Spatiotemporal modeling, on the contrary, could separate these overlapping activities (compare Figs. 4.2.1.2C and 2D). The model reproduced the data very closely (Figs. 4.2.1.2D and 4.2.1.2E), accounting for 88.7% of data variance. Run-to-run variability accounted for 5.1% of data variance in this patient.

Figs. 4.2.1.3 and 4.2.1.4 show the data and results of Patient #2, presented like Patient #1. As can be seen from Fig. 4.2.1.3B, both sources also were located in postcentral gyrus, about 5 mm beneath the cortical surface. Similar to Fig. 4.2.1.1B sources are depicted as surface projections in Fig. 4.2.1.3B. The source locations agreed with the results of cortical stimulations. A pure sensory response in the hand was produced by stimulation of the electrode next to the N20-P30 source. A mixed motor/sensory response in the hand could be elicited by stimulating the electrode next to the P25-N35 source (Fig. 4.2.1.3B). The distance from central sulcus was 10 mm for the surface projection of the N20-P30 source, and 8 mm for the surface projection of the P25-N35 source. In analogy to

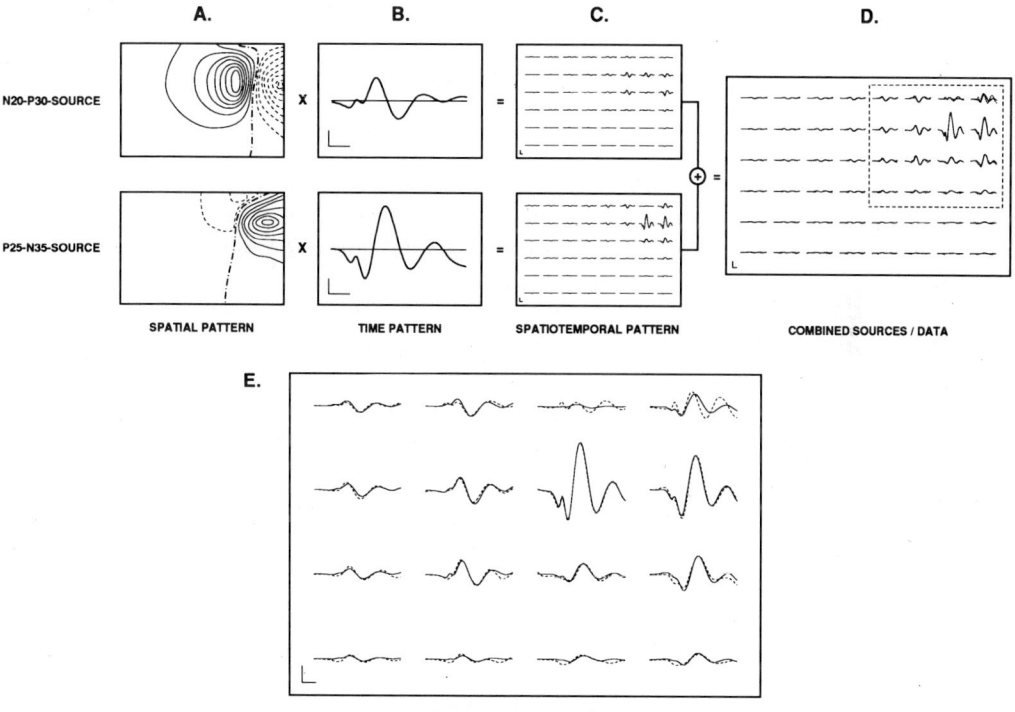

Fig. 4.2.1.4. Patient #2. Spatiotemporal structure of the somatosensory evoked response on ECoG. **A.** Spatial patterns of the individual sources shown as isopotential maps (10% isocontour lines; maximum of each map = 1 μV; polarities: positive (+) solid, negative (–) dashed). **B.** Time patterns of the individual sources. **C.** Spatiotemporal patterns of the individual sources, which is (A) times (B). **D.** Superposition of the combined spatiotemporal pattern of the two sources (solid) and the original data (dashed). **E.** Enlargement of region inside dotted rectangle in (D), showing superposition of combined spatiotemporal pattern (solid) and the original data (dashed). Calibration: horizontal = 10 msec, time scale begins at stimulus onset; vertical = 10 μV

Patient #1, the N20-P30 source was located lateral inferior to the P25-N35 source. The distance between the two sources was 11 mm. The time patterns of both sources showed peak latencies at 21.1 and 32.0 msec for the N20-P30 source, and at 25.8 and 37.5 msec for the P25-N35 source (Fig. 4.2.1.3C). In contrast to Patient #1, the time activities of both sources, especially that of the P25-N35 source, showed early peaks at about 15 msec in this patient (compare Figs. 4.2.1.1C and 4.2.1.3C). Similar to Patient #1, the time activities at 25 and 35 msec overlapped considerably (Fig. 4.2.1.3C).

Fig. 4.2.1.4 shows the spatiotemporal structure of the evoked response in this patient. Similar to the findings in Patient #1, the N20-P30 source was oriented in the anterior posterior direction and showed a tangential spatial potential pattern with a phase reversal across central sulcus (compare Figs. 4.2.1.2A and 4.2.1.4A, top rows). The P25-N35 source, on the contrary, showed a primarily radial pattern which was different from the findings in Patient #1 (compare Figs. 4.2.1.2A and 4.2.1.4A, bottom rows). The spatiotemporal patterns of the individual sources (Fig. 4.2.1.4C) were obtained by multiplying the corresponding spatial (Fig. 4.2.1.4A) and time patterns (Fig. 4.2.1.4B). In this patient, the contribution of the P25-N35 source to the evoked response was large compared to the contribution of the N20-P30 source, which was different from Patient #1. Again, the spatiotemporal patterns of the two sources showed considerable overlap but could probably be more easily identified from visual inspection of the raw data than in Patient #1 (compare Figs. 4.2.1.4C and 4.2.1.4D). Finally, the individual spatiotemporal patterns (Fig. 4.2.1.4C) were summed up to form the combined spatiotemporal pattern of the two sources (Fig. 4.2.1.4D) which reproduced the data very closely and explained 89.5% of the data variance. Run-to-run variability accounted for 6.3% of data variance in this patient.

TABLE 4.1.1. Patient data

Patient	Age[1]	Sex	Seizure origin	Seizure type[2]	Lesion
#1	12	M	Left frontal	CPS	Left frontal and temporal postencephalitic scar
#2	30	F	Left temporal	CPS	Left temporal scar
#3	28	M	Left frontal	SPS, CPS	Non-specific
#4	22	F	Right frontal	SPS	Right frontal AVM
#5	13	M	Right frontal	CPS	Right frontal glioma
#6	16	F	Right parietal	SPS	Non-specific

[1] Age = years
[2] CPS = complex partial seizures, SPS = simple partial seizures

4.2.2. Median Nerve Somatosensory Evoked Potentials on Scalp-EEG

4.2.2.1. Data

Median nerve SEPs were reproducible within subjects showing a run-to-run variability of 6-8%. There was some intersubject variability concerning wave morphology (compare Figs. 4.2.2.1B and 4.2.2.3B). Peak latencies were similar across subjects.

Median nerve SEPs showed prominent peaks at latencies of about 20 and 30 msec. At the frontal electrodes positive-negative waveforms were recordable, at the parietal electrodes negative-positive waveforms could be recorded. Corresponding to their polarity and latency, these waveforms have been termed P20-N30 components (frontal electrodes) and N20-P30 components (parietal electrodes) [4, 7, 117]. For sake of simplicity, we will refer to these components as N20 and P30 components in the following. Both N20 and P30, showed a phase reversal between frontal and parietal electrodes which approximately followed central sulcus. For Subject #1, these waveforms were clearly seen at the frontal superior and at the parietal inferior electrodes (see Fig. 4.2.2.1.B: frontal: column 5, rows 1-4; column 6, rows 1-4; parietal: column 1, rows 5-6; column 2, rows 5-6; columns numbered from left to right; rows numbered from top to bottom). For Subject #2, these waveforms could be best identified at the frontal superior electrodes (see Fig. 4.2.2.3.B: column 1, rows 1-4; column 2, rows 1-4).

In addition, positive peaks could be recorded at about 25 msec and negative peaks at about 35 msec. According to their polarity and latency, these components have been termed P25 and N35 components [4, 7, 117]. These components did not show a phase reversal across central sulcus, but instead showed single electropositivities and electronegativities, respectively. For Subject #1, these waveforms were clearly seen at the central electrodes (see Fig. 4.2.2.1.B: column 2, rows 2-4; column 3, rows 2-4; column 4, rows 2-4). For Subject #2, these waveforms could be best identified at the parietal electrodes (see Fig. 4.2.2.3.B: column 4, rows 3-5; column 5, rows 2-6; column 6, rows 2-6).

4.2.2.2. Number of Sources

First, we applied principal component analysis (PCA) to determine the number of underlying covarying processing. Two principal components accounted for 91.1% (range: 88.7-94.0%) and three principal components for 95.3% (range: 94.9-98.8%) of the data variance. Adding a third principal component, therefore, did not significantly increase the variance accounted for above noise. Thus, PCA indicated that there were two significant covarying parts underlying the system variance.

In a second step, we applied spatiotemporal modeling with two and three dipoles. Two sources accounted for an average of 87.7% of the data variance (range: 86.1-89.2%). Adding a third source explained less variance than run-to-run variability. We therefore concluded that two sources were underlying early SEPs recorded on scalp-EEG and used a model with two dipoles for further analysis.

4.2.2.3. Results of Spatiotemporal Modeling

Both sources had peak latencies which corresponded closely to the N20-P30 and P25-N35 peaks (peak latencies: source 1: 21 msec (range: 19-22 msec) and 31 msec (range: 30-33 msec); source 2: 26 msec (range: 24-27 msec) and 36 msec (range: 35-38 msec)). Therefore, source 1 will be referred to as N20-P30 source, and source 2 as P25-N35 source in the following.

The surface projection of both sources (radial projection of the three-dimensional intracerebral location of the source from the center of the sphere onto the surface of the scalp) was near C3/C4 in all subjects. The mean distance between the surface projection of N20-P30 source and C3/C4 was 9 mm (range: 6-11 mm), and 12 mm between the surface projection of the P25-N35 source and C3/C4 (range: 9-14 mm). The mean distance between the two sources was 12 mm (range: 8-14 mm). The N20-P30 source was consistently located deeper than the P25-N35 source in all subjects (mean depth of the N20-P30 source: 10 mm below the cortical surface, range: 7-13 mm; 23 mm below the scalp, range: 20-26 mm; depth of the P25-N35 source: 4 mm below the cortical surface, range: 2-5 mm; 17 mm below the scalp, range: 15-18 mm). There was no systematic difference in the medio-lateral or anterior-posterior location between the two sources.

In all subjects, the N20-P30 source consisted of a primarily tangential dipole oriented in the anterior-posterior direction resulting in an electric potential pattern which reversed polarity between the frontal and parietal electrodes. The phase reversal line approximately followed central fissure. The P25-N35 source consisted of a radial dipole resulting in an electric potential pattern with primarily a single electropositivity (P25) and electronegativity (N35), respectively.

The time patterns and the spatiotemporal potential patterns of both sources showed considerable overlap especially for activity after 20 msec. Thus, the contribution of each source to the evoked response could not be readily identified from visual inspection of the raw data.

4.2.2.4. Results of Selected Subjects

In the following, we present the results of some selected subjects.

The results for Subject #1 are shown in Figs. 4.2.2.1 and 4.2.2.2. In Fig. 4.2.2.1A the EEG recording matrix is shown in schematic form. The N20 and P30 components could be seen most clearly at the frontal and parietal electrodes, whereas the P25 and N35 components were best seen at the central electrodes (Fig. 4.2.2.1B; see also Section 4.2.2.1.).

Figs. 4.2.2.1C-E show the results of spatiotemporal modeling. In Figs. 4.2.2.1C and 4.2.2.1D the surface projections and the isopotential patterns of the sources are depicted. Both sources were close to the C4 position of the International 10-20 System (distance between the surface projection of the

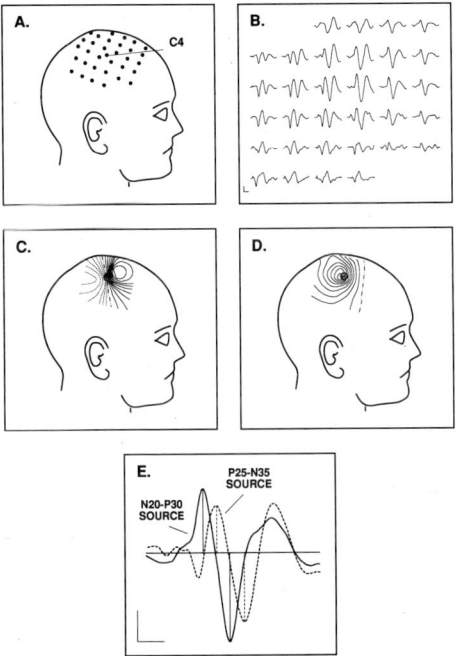

Fig. 4.2.2.1. Subject #1. **A.** Scalp-EEG was recorded simultaneously from a rectangular array of 32 electrodes which were arranged in a 6 times 6 matrix centered at C4. **B.** SEPs on scalp-EEG. **C., D.** Surface projections (indicated by circled dots) and isopotential maps of the two sources in schematic form. Isopotential maps: 10% isocontour lines; maximum of each map = 1 μV; polarities: positive (+) solid, negative (−) dashed. **E.** Superimposed time activities of the two sources show considerable overlap. Vertical lines indicate peak latencies. Calibration: horizontal = 10 msec, time scale begins at stimulus onset; vertical = 1 μV

N20-P30 source and C4 = 11 mm; distance between surface projection of the P25-N35 source and C4 = 14 mm). The distance between the two sources was 8 mm. The N20-P30 source was located slightly deeper than the P25-N35 source (depth of the N20-P30 source: 7 mm below the cortical surface and 20 mm below the scalp; depth of the P25-N35 source: 4 mm below the cortical surface and 17 mm below the scalp). The N20-P30 source consisted of a tangential dipole oriented primarily in the anterior-posterior direction resulting in a spatial potential pattern which reversed polarity between the frontal and parietal electrodes. The phase reversal line approximately followed central sulcus (Figs. 4.2.2.1C and 4.2.2.2A, top row). The P25-N35 source consisted of a radial dipole producing a single electropositivity (P25) and electronegativity (N35) (Figs. 4.2.2.1D and 4.2.2.2A, bottom row). The time activities had peak latencies at 21.3 and 30.6 msec for the N20-P30 source, and at 26.0 and 36.0 msec for the P25-N35 source, and showed considerable overlap especially for activity after 20 msec (Fig. 4.2.2.1E). Furthermore, some minor activity was found for both sources around 15 msec before the first cortical activity.

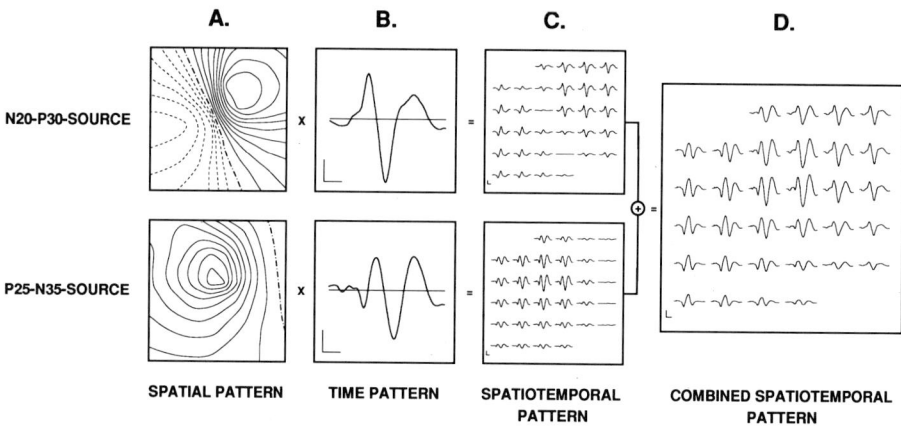

Fig. 4.2.2.2. Subject #1. Spatiotemporal structure of the somatosensory evoked response on scalp-EEG. **A.** Spatial patterns of the individual sources shown as isopotential maps (10% isocontour lines; maximum of each map = 1 μV; polarities: positive (+) solid, negative (–) dashed). **B.** Time patterns of the individual sources. **C.** Spatiotemporal patterns of the individual sources, which is (A) times (B). **D.** Combined spatiotemporal pattern of the two sources which could reproduce the data very closely accounting for 86.1% of the data variance (compare Fig. 4.2.2.1B). Calibration: horizontal = 10 msec, time scale begins at stimulus onset; vertical = 1 μV

The results of multiple dipole modeling and the spatiotemporal structure of the SEPs are shown in Fig. 4.2.2.2. Each source was uniquely represented by a spatial potential pattern (Fig. 4.2.2.2A) and a time pattern (Fig. 4.2.2.2B). The spatiotemporal potential patterns (Fig. 4.2.2.2C) of the individual sources were obtained by multiplying the spatial patterns (Fig. 4.2.2.2A) with the time patterns (Fig. 4.2.2.2B) and showed considerable overlap. It should be noted that the spatiotemporal structures of the individual sources could not be readily identified from visual inspection of the raw data (compare Figs. 4.2.2.1B, 4.2.2.2C, and 4.2.2.2D). The contribution of the individual sources to the evoked response was about equal. The combined spatiotemporal pattern (Fig. 4.2.2.2D) of the two sources was calculated by summing up the individual spatiotemporal patterns (Fig. 4.2.2.2D) according to the law of superposition and could reproduce the data closely, accounting for 86.1% of the data variance (compare Figs. 4.2.2.1B and 4.2.2.2D).

Figs. 4.2.2.3 and 4.2.2.4 show data and results for subject #2, presented in the same format as subject #1. Figs. 4.2.2.3A and 4.2.2.3B show EEG recording matrix and data for this subject. Whereas P25 and N35 components could be seen very clearly at the parietal electrodes, N20 and P30 components were difficult to discern at the frontal electrodes and could not be identified at the parietal electrodes at all. These findings were different from Subject #1 (compare Figs. 4.2.2.1B and 4.2.2.3B, see also Section 4.2.2.1).

The surface projections, isopotential maps, and time activities of the sources are shown in Figs. 4.2.2.3C-E. Both sources were close to the C3

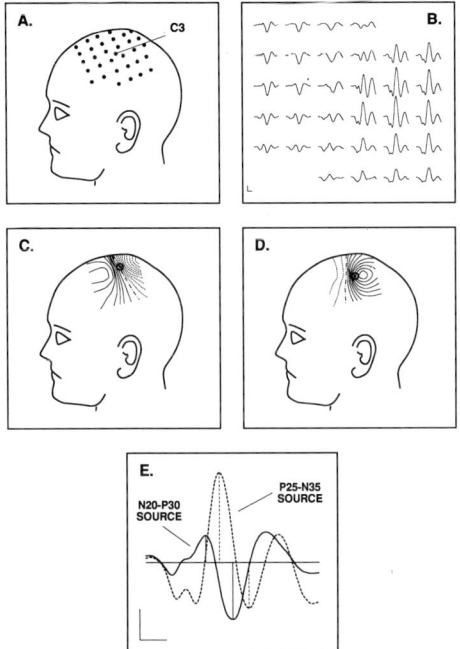

Fig. 4.2.2.3. Subject #2. **A.** Scalp-EEG was recorded simultaneously from a rectangular array of 32 electrodes which were arranged in a 6 times 6 matrix centered at C3. **B.** SEPs on scalp-EEG. **C., D.** Surface projections (indicated by circled dots) and isopotential maps of the two sources in schematic form. Isopotential maps: 10% isocontour lines; maximum of each map = 1 μV; polarities: positive (+) solid, negative (–) dashed. **E.** Superimposed time activities of the two sources show considerable overlap. Vertical lines indicate peak latencies. Calibration: horizontal = 10 msec, time scale begins at stimulus onset; vertical = 1 μV

position of the International 10-20 System (distance between the surface projection of the N20-P30 source and C3 = 9 mm; distance between the surface projection of the P25-N35 source and C3 = 12 mm). The distance between the two sources was 13 mm. In contrast to Subject #1, the depth estimates of the sources were considerably different with the N20-P30 source located significantly deeper compared to the P25-N35 source (depth of the N20-P30 source: 10 mm below the cortical surface and 23 mm below the scalp; depth of the P25-N35 source: 4 mm below the cortical surface and 17 mm below the scalp). Similar to Subject #1, the N20-P30 source was a tangential dipole arranged in the anterior-posterior direction and had a spatial potential pattern with a phase reversal between frontal and parietal electrodes (compare Figs. 4.2.2.3C and 4.2.2.4A, top row). The P25-N35 source was a radial dipole producing a single electropositivity (P25) and electronegativity (N35), respectively (Figs. 4.2.2.3D and 4.2.2.4A, bottom row). The time patterns had peak latencies at 21.8 and 32.0 msec for the N20-P30 source, and at 26.9 and 38.0 msec for the P25-N35 source. These time activities showed considerable overlap especially for activity after 20 msec (Fig. 4.2.2.1E). In contrast to Subject #1, there was considerable

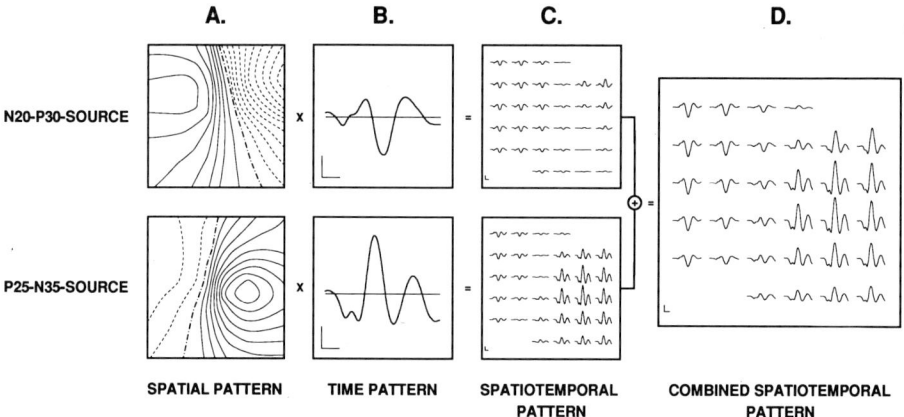

A. Spatial patterns of the individual sources shown as isopotential maps

N20-P30-SOURCE

P25-N35-SOURCE

SPATIAL PATTERN　　TIME PATTERN　　SPATIOTEMPORAL PATTERN　　COMBINED SPATIOTEMPORAL PATTERN

Fig. 4.2.2.4. Subject #2. Spatiotemporal structure of the somatosensory evoked response on scalp-EEG. **A.** Spatial patterns of the individual sources shown as isopotential maps (10% isocontour lines; maximum of each map = 1 μV; polarities: positive (+) solid, negative (–) dashed). **B.** Time patterns of the individual sources. Calibration: horizontal = 10 msec, time scale begins at stimulus onset; vertical = 1 μV. **C.** Spatiotemporal patterns of the individual sources, which is (A) times (B). **D.** Combined spatiotemporal pattern of the two sources which could reproduce the data very closely accounting for 89.2% of the variance (compare Fig. 4.2.2.3B). Calibration: horizontal = 10 msec, time scale begins at stimulus onset; vertical = 1 μV

activity of both sources around 15 msec before the first cortical activity, especially for the P25-N35 source.

　　The results of spatiotemporal modeling for subject #2 are shown in Fig. 4.2.2.4. Again, the spatiotemporal patterns of the individual sources (Fig. 4.2.2.4C) were calculated by multiplying the spatial potential patterns (Fig. 4.2.2.4A) with the corresponding time activity patterns (Fig. 4.2.2.4B). It should be noted that the contribution of the P25-N35 source largely dominated the evoked response, whereas the contribution of the N20-P30 source was only minor. This is in contrast to Subject #1, where both sources showed a significant and almost equal contribution to the SEPs (compare Figs. 4.2.2.2C and 4.2.2.4C). Finally, the combined sources could reproduce the spatiotemporal pattern of the data very closely accounting for 89.2% of the data variance (compare Figs. 4.2.2.3B and 4.2.2.4D).

4.2.3. Median Nerve Somatosensory Evoked Fields on Magnetoencephalography

4.2.3.1. Data

　　Median nerve somatosensory evoked magnetic fields (SEFs) were reproducible within subjects showing a run-to-run variability of 5-8%. Furthermore, SEF waveforms and peak latencies were similar across subjects (compare Figs. 4.2.3.1B and 4.2.3.3B).

　　Median nerve SEFs showed prominent peaks at latencies of about 20 and 30 msec. According to their latency, these components will be referred to as SEF_{20} and SEF_{30} components in the following (SEF = somatosensory evoked magnetic field). Whereas the magnetic field for the SEF_{20} component emerged from the head superior and re-entered the head inferior, the magnetic field for the SEF_{30} component was of opposite direction and emerged from the head inferior and re-entered the head superior. Thus, both SEF_{20} and SEF_{30} showed magnetic field patterns with a phase reversal between the upper and lower recording positions.

Additional peaks could not be readily identified especially for Subject #1. However, closer inspection of the data for Subject #2 showed additional small inflections superimposed on the slope between the SEF_{20} and SEF_{30} components in the 20-30 msec range, and after the SEF_{30} component in the 30-40 msec range. According to their latency, these components will be referred to as SEF_{25} and SEF_{35} components. A clear and consistent magnetic field pattern could not be identified for these components as they were hidden in the SEF_{20}-SEF_{30} sequence.

4.2.3.2. Number of Sources

Principal component analysis (PCA) indicated two significant covarying parts underlying the system variance as two principal components accounted for an average of 90.6% of the data variance (range: 85.4-96.1%), and three principal components accounted for an average of 92.7% of the variance (range: 88.1-97.8%). Using multiple dipole modeling, two sources accounted for 86.4% of the data variance (range: 81.2-92.6%). Adding a third source did not significantly increase the variance accounted for and was smaller than run-to-run noise. Thus, it was concluded that two dipolar sources were sufficient to model the first 60 msec of the SEFs in all 9 subjects.

4.2.3.3. Results of Spatiotemporal Modeling

Both sources showed biphasic time activities. Source 1 had peak latencies at 21 msec (range: 19-22 msec) and 31 msec (range: 29-32 msec) and therefore will be referred to as SEF_{20}-SEF_{30} source. Source 2 showed peak latencies at 24 msec (range: 22-26 msec) and 34 msec (range: 32-36 msec) and thus was termed SEF_{25}-SEF_{35} source.

The surface projection of both sources (radial projection of the three-dimensional intracerebral location of the source from the center of the sphere onto the surface of the scalp) was near C3/C4 in all subjects. The mean distance between the surface projection of the SEF_{20}-SEF_{30} source and C3/C4 was 9 mm (range: 4-14 mm), and 13 mm between the surface projection of the SEF_{25}-SEF_{35} source and C3/C4 (range: 8-17 mm). The mean distance between the two sources was 9 mm (range: 6-13 mm). The SEF_{20}-SEF_{30} source was consistently located deeper than the SEF_{25}-SEF_{35} source in all subjects (mean depth of the SEF_{20}-SEF_{30} source: 9 mm below the cortical surface, range: 6-12 mm; 22 mm below the scalp, range: 19-25 mm; depth of the SEF_{25}-SEF_{35} source: 3 mm below the cortical surface, range: 1-5 mm; 16 mm below the scalp, range: 14-18 mm). There was no systematic difference in the medio-lateral or anterior-posterior location between the two sources.

In all subjects, the SEF_{20}-SEF_{30} source was oriented primarily in the anterior-posterior direction resulting in a magnetic field pattern which reversed polarity between the upper and lower recording positions emerging from the head medial superior and re-entering the head lateral inferior. The SEF_{25}-SEF_{35} source showed no consistent orientation across subjects.

The spatiotemporal field patterns of both sources showed considerable overlap both in space and time in all subjects.

The variance explained by the SEF_{20}-SEF_{30} source (mean: 69.2%, range: 62.1-71.2%) was considerably larger than that explained by the SEF_{25}-SEF_{35} source (mean: 17.3%, range: 10.1-22.8%).

4.2.3.4. Results of Selected Subjects

In the following we present the results of some selected subjects. Measurement matrix and SEFs of subject #1 are shown in Figs. 4.2.3.1A-B.

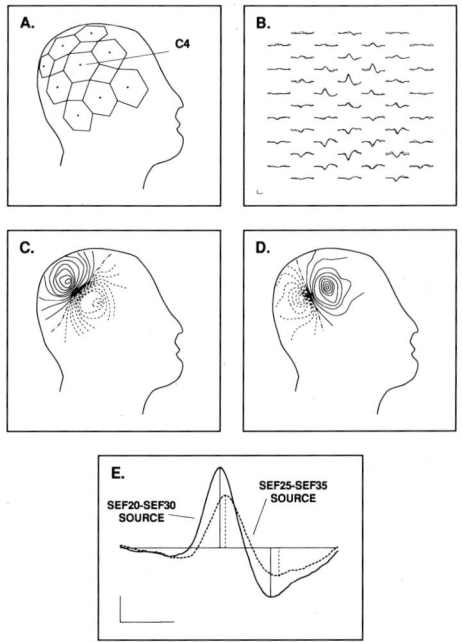

Fig. 4.2.3.1. Subject #1. **A.** Measurement grid. Measurements were obtained from 9 overlapping probe positions yielding 45 distinct measurement points. **B.** Median nerve SEFs. Two runs of 500 trials each are superimposed to show reproducibility. Calibration: horizontal = 10 msec, time scale begins at 4 msec after stimulus onset; vertical = 100 fT. **C., D.** Surface projections (indicated by crosses) and isofield maps of the two sources in schematic form. Isofield maps: 10% isocontour lines; maximum of each map = 1 fT; solid lines: magnetic field emerging from the head; dashed lines: magnetic field re-entering the head. **E.** Time activities of the two sources show considerable overlap. Calibration: horizontal = 10 msec, time scale begins at 4 msec after stimulus onset; vertical = 50 fT

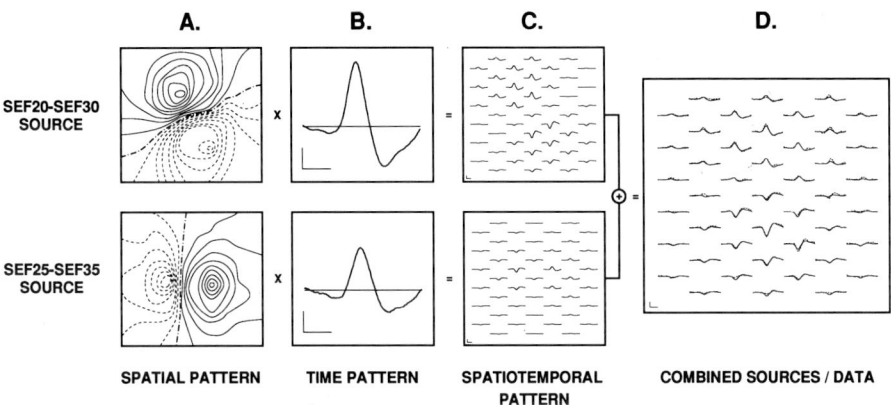

A. **B.** **C.** **D.**

SEF20-SEF30
SOURCE

SEF25-SEF35
SOURCE

SPATIAL PATTERN   TIME PATTERN   SPATIOTEMPORAL   COMBINED SOURCES / DATA
PATTERN

Fig. 4.2.3.2. Subject #1. Spatiotemporal structure of the somatosensory evoked response on MEG. **A.** Spatial magnetic field patterns of the individual sources presented as isofield maps (10% isocontour lines; maximum of each map = 1 fT; solid lines: magnetic field emerging from the head; dashed lines: magnetic field re-entering the head). **B.** Time patterns of the individual sources. Calibration: horizontal = 10 msec, time scale begins at 4 msec after stimulus onset; vertical = 50 fT. **C.** Spatiotemporal field patterns of the individual sources, which is (A) times (B). **D.** The combined spatiotemporal pattern of the two sources (solid lines) was calculated by summing up the individual spatiotemporal patterns and could reproduce the data (dashed lines) closely, accounting for 83.6% of the variance. Calibration: horizontal = 10 msec, time scale begins at 4 msec after stimulus onset; vertical = 100 fT

The SEF_{20}-SEF_{30} source was located deeper than the SEF_{25}-SEF_{35} source (depth of the SEF_{20}-SEF_{30} source: 10 mm below the cortical surface and 23 mm below the scalp, respectively; depth of the SEF_{25}-SEF_{35} source: 1 mm below the cortical surface and 14 mm below the scalp, respectively). Surface projections and isofield maps of the individual sources are shown schematically in Figs. 4.2.3.1C-D. The distance between the surface projections of the sources and C4 was 4 mm for the SEF_{20}-SEF_{30} source, and 8 mm for the SEF_{25}-SEF_{35} source. The distance between the two sources was 8 mm. The SEF_{20}-SEF_{30} source was oriented primarily in the anterior-posterior direction resulting in a spatial magnetic field pattern which reversed polarity between the upper and lower recording positions. Note that the phase reversal was slightly tilted from the horizontal plane and thus was approximately perpendicular to central sulcus. The SEF_{25}-SEF_{35} source was oriented primarily in the superior-inferior direction resulting in a spatial field pattern which reversed polarity between the anterior and posterior recording positions. The time patterns of both sources were biphasic with peak latencies of 21.9 and 31.6 msec for the SEF_{20}-SEF_{30} source, and of 23.3 and 33.0 msec for the SEF_{25}-SEF_{35} source, and showed considerable overlap (Fig. 4.2.3.1E).

The spatiotemporal structure of the somatosensory evoked response for Subject #1 is presented in Fig. 4.2.3.2. Each source was uniquely represented by a spatial field pattern (Fig. 4.2.3.2A) and a time pattern (Fig. 4.2.3.2B). The spatiotemporal field patterns (Fig. 4.2.3.2C) of the individual sources

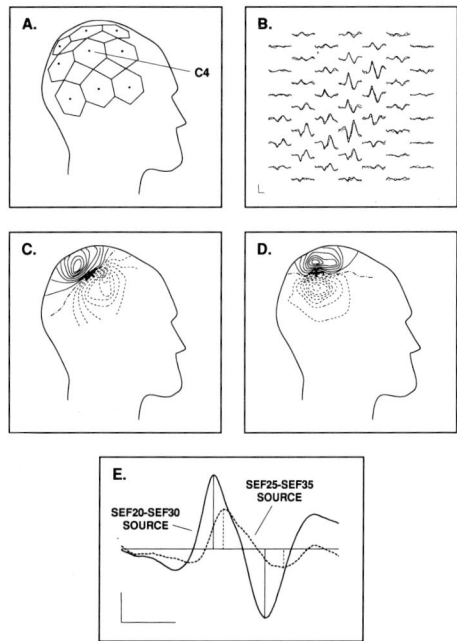

Fig. 4.2.3.3. Subject #2. **A.** Measurement grid. Measurements were obtained from 9 overlapping probe positions yielding 45 distinct measurement points. **B.** Median nerve SEFs. Two runs of 500 trials each are superimposed to show reproducibility. Calibration: horizontal = 10 msec, time scale begins at 4 msec after stimulus onset; vertical = 100 fT. **C., D.** Surface projections (indicated by crosses) and isofield maps of the two sources in schematic form. Isofield maps: 10% isocontour lines; maximum of each map = 1 fT; solid lines: magnetic field emerging from the head; dashed lines: magnetic field re-entering the head. **E.** Time activities of the two sources show considerable overlap. Calibration: horizontal = 10 msec, time scale begins at 4 msec after stimulus onset; vertical = 50 fT

were obtained by multiplying the spatial patterns with the corresponding time patterns and showed considerable overlap both in space and time. Thus, the magnetic field contributions of the individual sources could not be readily identified by mere visual inspection of the raw data (compare Figs. 4.2.3.1B, 4.2.3.2C, and 4.2.3.2D). Furthermore, it should be noted that the contribution of the SEF_{20}-SEF_{30} source to the evoked response was much larger than that of the SEF_{25}-SEF_{35} source (compare Fig. 4.2.3.2C top and bottom). The combined spatiotemporal pattern (Fig. 4.2.3.2D) of the two sources was calculated by summing up the individual spatiotemporal patterns and could reproduce the data closely, accounting for 83.6% of the variance (Fig. 4.2.3.2D).

Figs. 4.2.3.3 and 4.2.3.4 show data and results for subject #2, presented in the same format as subject #1. Again, the SEF_{20}-SEF_{30} source was located deeper than the SEF_{25}-SEF_{35} source (depth of the SEF_{20}-SEF_{30} source: 9 mm below the cortical surface and 22 mm below the scalp, respectively; depth of the SEF_{25}-SEF_{35} source: 3 mm below the cortical surface and 16 mm below

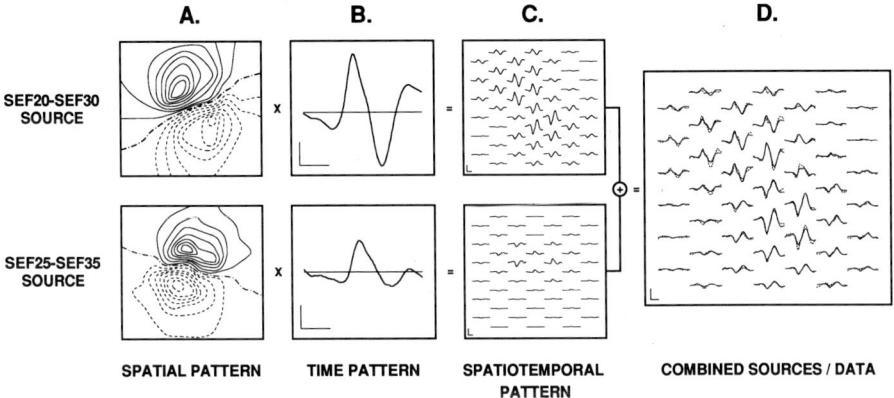

A.   **B.**   **C.**   **D.**

SEF20-SEF30 SOURCE

SEF25-SEF35 SOURCE

SPATIAL PATTERN   TIME PATTERN   SPATIOTEMPORAL PATTERN   COMBINED SOURCES / DATA

Fig. 4.2.3.4. Subject #2. Spatiotemporal structure of the somatosensory evoked response on MEG. **A.** Spatial magnetic field patterns of the individual sources presented as isofield maps (10% isocontour lines; maximum of each map = 1 fT; solid lines: magnetic field emerging from the head; dashed lines: magnetic field re-entering the head). **B.** Time patterns of the individual sources. Calibration: horizontal = 10 msec, time scale begins at 4 msec after stimulus onset; vertical = 50 fT. **C.** Spatiotemporal field patterns of the individual sources, which is (A) times (B). **D.** The combined spatiotemporal pattern of the two sources (solid lines) was calculated by summing up the individual spatiotemporal patterns and could reproduce the data (dashed lines) closely, accounting for 81.9% of the variance. Calibration: horizontal = 10 msec, time scale begins at 4 msec after stimulus onset; vertical = 100 fT

the scalp, respectively). The distance between C4 and the surface projection of the SEF_{20}-SEF_{30} source was 6 mm, and was 16 mm between C4 and the surface projection of the SEF_{25}-SEF_{35} source. The distance between the two sources was 14 mm (Figs. 4.2.3.3C and 4.2.3.3D). As in subject #1, the SEF_{20}-SEF_{30} source was oriented primarily in the anterior-posterior direction with a phase reversal perpendicular to central sulcus. Unlike in subject #1, the SEF_{25}-SEF_{35} source also was oriented primarily in the anterior-posterior direction with a more horizontal phase reversal compared to the SEF_{20}-SEF_{30} source (compare Figs. 4.2.3.1C and 4.2.3.3C, Figs. 4.2.3.1D and 4.2.3.3D, and Figs. 4.2.3.2A and 4.2.3.4A). The time patterns of both sources again were biphasic (peak latencies 20.9 and 30.3 msec for the SEF_{20}-SEF_{30} source; 22.7 and 34.1 msec for the SEF_{25}-SEF_{35} source) and showed considerable overlap (Fig. 4.2.3.3E). The spatiotemporal structure of the evoked response for Subject #2 is shown in Fig. 4.2.3.4. Analogous to Subject #1, the spatial patterns (Fig. 4.2.3.4A) were multiplied with the time patterns (Fig. 4.2.3.4B) to achieve the spatiotemporal magnetic field contributions of the individual sources (Fig. 4.2.3.4C). These individual spatiotemporal patterns were summed up form the combined spatiotemporal pattern which could reproduce the data very closely accounting for 81.9% of the variance (Fig. 4.2.3.4D).

4.2.4. Comparison of ECoG, Scalp-EEG, and MEG

4.2.4.1. Comparison of the Raw Data for the Subjects as a Group

It should be mentioned that we could not record somatosensory evoked responses on all subjects with all three different techniques, i.e. ECoG, scalp-EEG, and MEG. Thus, these general results are based on comparisons of subject groups, rather than on intraindividual comparisons. Thus, our results are affected by intersubject variability of evoked responses [8, 32, 142, 346, 378]. Nevertheless, we had the opportunity to perform investigations with different techniques in some subjects. The results on these selected subjects will be presented in Sections 4.2.4.3 and 4.2.4.4, respectively.

On comparison of the raw data, SEPs on ECoG showed a much larger amplitude than those on scalp-EEG. Thus, the N20-P30 amplitude was an average of 51.4 μV on ECoG, and 4.6 μV on scalp-EEG, which represents an attenuation factor of 11.2. It should be noted that especially on ECoG there was a wide variability of amplitude. The N20-P30 amplitude had a range from 35 μV to 125 μV. Furthermore, the responses on ECoG were extremely focal with a rapid fall off over space. SEPs on scalp-EEG, on the contrary, had a widespread distribution (Fig. 4.2.4.1). Concerning wave morphology, there was variability between subjects within each recording technique. Thus, it does not make much sense to compare SEP wave morphologies for different subjects recorded with different techniques. The issue of wave morphology will be discussed in Section 4.2.4.3, where SEPs were recorded on ECoG and scalp-EEG in the same subject. MEG cannot be compared directly to electrical recording techniques concerning amplitude. Concerning spatial distribution of the evoked response, MEG was more focal than scalp-EEG, but more widespread than ECoG (compare Figs. 4.2.4.1 and 4.2.4.4).

4.2.4.2. Comparison of Spatiotemporal Modeling for the Subjects as a Group

Spatiotemporal modeling basically yielded identical results with three different recording techniques (ECoG, scalp-EEG, and MEG), namely that the majority of the human somatosensory evoked response is generated by two neuronal sources which are located in the posterior wall of central sulcus and the anterior crown of postcentral gyrus, respectively. On all techniques, there was considerable overlap of the spatiotemporal patterns generated by the two sources. The spatiotemporal structures of the individual sources were similar across techniques. One source generated the electric N20 and P30 components as well as the magnetic SEF_{20} and SEF_{30} components. We will refer to this source as N20-P30 source and SEF_{20}-SEF_{30} source, respectively. The other source generated the electric P25 and N35 components as well as the magnetic SEF_{25} and SEF_{35} components and will be referred to as P25-N35 source and as SEF_{25}-SEF_{35} source, respectively.

On ECoG, both sources were located in postcentral gyrus within 10 mm of central sulcus (mean distance = 6 mm). On both scalp-EEG and MEG, surface projections of the two sources were within 17 mm of the C3/C4 position of the International 10-20 System which in turn is within 15 mm of central sulcus [173, 174, 180, 331]. The mean distance of the surface projections of the sources from C3/C4 was 11 mm on scalp-EEG, and 12 mm on MEG. Thus, it can be concluded that the localization estimates concurred within the accuracy limits of the different methods.

The average depth estimates below the cortical surface for the N20-P30 and SEF_{20}-SEF_{30} sources were similar across methods, i.e. 8 mm on ECoG, 10 mm on scalp-EEG, and 9 mm on MEG. The depth estimates for the P25-N35 and SEF_{25}-SEF_{35} sources were more variable across methods yielding estimates of 8 mm on ECoG, 4 mm on scalp-EEG, and 3 mm on MEG. Both on scalp-EEG and MEG, the depth estimates for the P25-N35 and SEF_{25}-SEF_{35} sources were more superficial compared to those for the N20-P30 and SEF_{20}-SEF_{30} sources in all subjects, whereas on ECoG, there was no consistent difference of depth estimates for the N20-P30 sources and the P25-N35 sources.

Concerning source orientation, the N20-P30 source consisted of a tangential dipole pointing in the anterior direction both on ECoG and scalp-EEG. Thus, the isopotential maps for this source were positive over the frontal (scalp-EEG) or precentral (ECoG) electrodes, and negative over the parietal (scalp-EEG) or postcentral (ECoG) electrodes with a phase reversal following approximately central sulcus. In agreement with these results, the magnetic SEF_{20}-SEF_{30} source also was a tangential dipole pointing in the anterior direction resulting in a magnetic field pattern which emerged from the head superior and re-entered the head inferior. The electric isopotential maps and the magnetic isofield maps were perpendicular to each other as expected. For this source, isopotential maps were most focal on ECoG showing a mean distance of 1.9 cm between the peak amplitudes. MEG had a more restricted pattern compared to scalp-EEG, as the distance between magnetic field maxima averaged 6.1 cm, whereas the distance between electric potential maxima averaged 10.3 cm. Thus, the results concerning the electric N20-P30 source and the magnetic SEF_{20}-SEF_{30} source obtained with the different techniques yielded confirmatory evidence to each other.

The orientations of the P25-N35 source and the corresponding SEF_{25}-SEF_{35} source were more variable across methods. On ECoG, the P25-N35 source showed variable orientations. In some patients the P25-N35 source was oriented more radially, in other patients it was oriented more tangentially. Contrarily, on scalp-EEG the P25-N35 source consistently showed a radial pattern. On MEG, finally, the equivalent SEF_{25}-SEF_{35} source showed inconsistent orientations across subjects. In this context, it should be mentioned that MEG only detects the tangential components of this primarily radial source and thus the results of MEG concerning the SEF_{25}-SEF_{35} source are not directly comparable to those of the P25-N35 source on ECoG and scalp-EEG.

Finally, the time activities and peak latencies of the individual sources showed good agreement across methods.

4.2.4.3. Comparison of ECoG and Scalp-EEG in a Selected Patient

In one patient, we had the opportunity to record both ECoG and scalp-EEG. This patient was a 30 year old female with medically refractory complex partial seizures. Bilateral depth recordings consistently pointed towards a left hemispheric seizure origin. Typical seizures showed initial involvement of speech, i.e. speech arrest. However, a temporal or frontal lobe seizure origin could not clearly be distinguished. Thus, subdural grids were implanted to map out essential cortex, specifically language-related cortex, and to definitely localize the seizure focus. Eventually, a left temporal seizure focus could be identified and a left anterior temporal lobectomy was performed. Histopathologic examination of the specimen revealed mesial temporal sclerosis. Concerning outcome, the patient experienced a significant (>90%) reduction in seizure frequency, and had no functional deficits induced by the operation.

Fig. 4.2.4.1 shows median nerve SEPs recorded from the cortical surface and from the scalp. Wave morphologies showed striking similarities between both recording techniques. Typical traces on ECoG and scalp-EEG were superimposed for comparison and showed almost identical waveforms. ECoG showed a restricted spatial distribution with a very rapid fall-off over space. In this context, it should be noted that ECoG electrodes were spaced 10 mm center-to-center, whereas EEG electrodes had a spacing of 30 mm. The N20-P30 amplitude on cortical surface recordings was

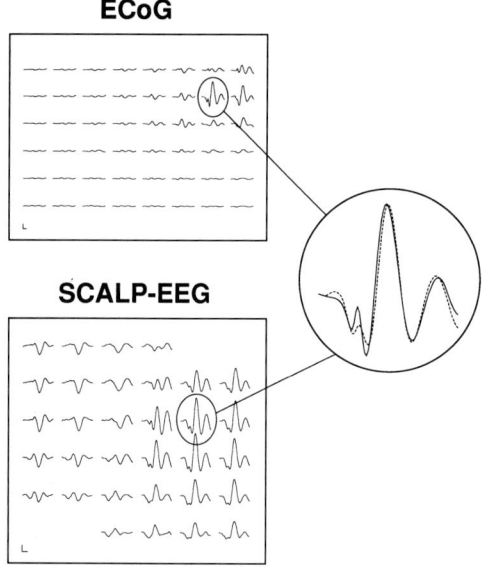

Fig. 4.2.4.1. Somatosensory evoked potentials recorded on electrocorticography (top) and scalp-EEG (bottom). Typical traces are superimposed in the circle on the left for comparison (ECoG: solid, scalp-EEG: dashed). Polarities: positive (+) up, negative (–) down. Calibration: horizontal = 10 msec, time scale begins at stimulus onset; vertical = 10 µV (ECoG) and 1 µV (scalp-EEG)

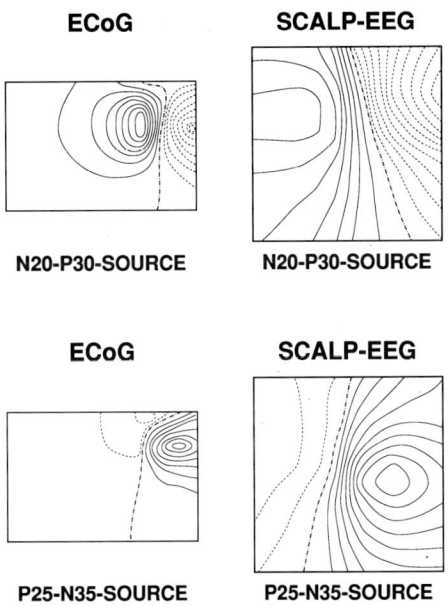

Fig. 4.2.4.2. Spatial potential patterns of the individual sources on electro-corticography (left) and scalp-EEG (right) shown as isopotential maps (10% isocontour lines; maximum of each map = 1 µV; polarities: positive (+) solid, negative (–) dashed). The N20-P30 source (top) consisted of a tangential dipole pointing in the anterior direction and resulted in a spatial potential pattern with a phase reversal between the anterior and posterior electrodes following approximately central sulcus. The P25-N35 source (bottom) was represented by a radial dipole and resulted in a spatial potential pattern with a single electropositivity. The similarities of the isopotential maps of the two sources obtained on ECoG and scalp-EEG is evident. It should be noted that the spatial potential pattern on ECoG was much more focal than on scalp-EEG. For the N20-P30 source, the distance between amplitude maxima was 1.9 cm on ECoG, and 11.6 cm on scalp-EEG

46.7 µV, that on scalp recordings 4.7 µV which corresponds to an attenuation factor of 9.9.

In a next step, we compared the results of spatiotemporal modeling obtained with the two techniques. We found that the majority of the somatosensory evoked response both on ECoG and scalp-EEG could be explained by two neuronal sources producing potentials overlapping both in space and time, namely a N20-P30 and a P25-N35 source. Both on ECoG and scalp-EEG, the N20-P30 source consisted of a tangential dipole pointing in the anterior direction which resulted in a spatial potential pattern with a phase reversal between the anterior and posterior electrodes. The P25-N35 source was represented by a primarily radial dipole resulting in a spatial potential pattern with a single electropositivity. The similarities of the isopotential maps of these two sources obtained on ECoG and scalp-EEG can be seen in Fig. 4.2.4.2. It should be noted that the spatial potential pattern on ECoG was much more focal than on scalp-EEG. For the N20-P30 source, the distance between amplitude maxima was 1.9 cm on ECoG, and

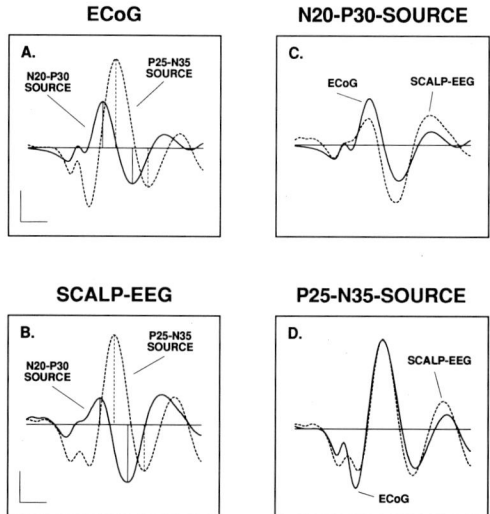

Fig. 4.2.4.3. A., B. Superimposed time patterns of the N20-P30 and the P25-N35 source are shown on the left for ECoG (A) and for scalp-EEG (B). Vertical lines indicate peak latencies. Polarities: positive (+) up, negative (−) down. Calibration: horizontal = 10 msec, time scale begins at stimulus onset; vertical = 10 µV (ECoG) and 1 µV (scalp-EEG). **C., D.** Superposition of the time activities obtained with the different techniques for the N20-P30 source (C) and for the P25-N35 source (D). The time patterns were very similar across techniques. ECoG: solid, scalp-EEG: dashed

11.6 cm on scalp-EEG. The results on depth estimates in this patient were different compared to the patients as a group. In this patient, the N20-P30 source was located more superficial on ECoG than on scalp-EEG, i.e. 4 mm on ECoG versus 10 mm on scalp-EEG. On the contrary, the depth estimates obtained for the P25-N35 source were similar, i.e. 6 mm on ECoG versus 4 mm on scalp-EEG.

The time activities of the two sources are displayed in Fig. 4.2.4.3. For both sources, waveforms and peak latencies were very similar on ECoG and scalp-EEG. This holds especially true for the P25-N35 source, where the time activities were almost identical on ECoG and on scalp-EEG. Thus, it can be concluded that spatiotemporal modeling separated out similar spatiotemporal patterns both on ECoG and scalp-EEG.

4.2.4.4. Comparison of Scalp-EEG and MEG in a Selected Subject

Scalp-EEG and MEG comparison is presented for a 31 year old male volunteer. Fig. 4.2.4.4 shows the raw data for this subject. The evoked responses looked different on scalp-EEG and MEG as can be seen from the left circle where typical traces are superimposed. Scalp-EEG looked more complicated as it showed additional peaks, namely prominent P25 peaks. This was expected because the P25 component is generated by a radial dipole which can be measured on scalp-EEG, whereas on MEG only its small tangential components can be detected.

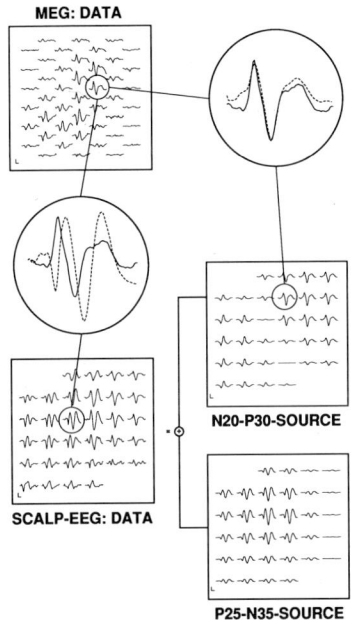

Fig. 4.2.4.4. Somatosensory evoked responses recorded on MEG and scalp-EEG. The raw data on MEG and scalp-EEG were different as can be seen from superposition of typical traces in the circle to the left. Scalp-EEG was decomposed into its tangential and radial components, i.e. potential contributions from the N20-P30 and P25-N35 components, respectively (lower right). Tangential scalp-EEG components (N20-P30 source) and MEG raw data showed similar waveforms as can be seen from superposition of typical traces in the right upper circle. MEG: solid, scalp-EEG: dashed. Calibration: horizontal = 10 msec, time scale begins at stimulus onset; vertical = 100 fT (MEG) and 1 μV (scalp-EEG)

We therefore decomposed the scalp-EEG data in their radial and tangential components. In order to achieve this, we utilized brain physiology, i.e. that the N20-P30 source generates primarily the tangential EEG components of the evoked response, whereas the P25-N35 source generates primarily its radial EEG components. We therefore applied spatiotemporal modeling and compared the potential contribution of the N20-P30 source, i.e. the tangential part of the evoked response on scalp-EEG to the MEG raw data. The results of this analysis are shown in Fig. 4.2.4.4. Two typical traces are enlarged for clarity and superimposed in the right upper circle. There was an excellent agreement between the tangential EEG components and the MEG raw data.

We then compared the results of spatiotemporal modeling obtained on scalp-EEG and MEG. Both techniques yielded two neuronal sources with considerable overlap both in space and time. The spatial patterns of the N20-P30 source on scalp-EEG and of the SEF_{20}-SEF_{30} source on MEG both showed tangential dipolar patterns (Fig. 4.2.4.5, top row). These patterns were orthogonal to each other and thus consistent with a tangential dipole in the posterior wall of central sulcus. The magnetic field pattern was more

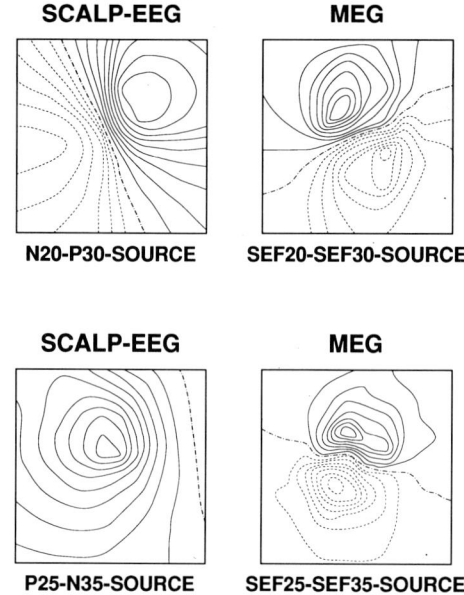

SCALP-EEG   **MEG**

N20-P30-SOURCE   **SEF20-SEF30-SOURCE**

SCALP-EEG   **MEG**

P25-N35-SOURCE   **SEF25-SEF35-SOURCE**

Fig. 4.2.4.5. Spatial field patterns of the individual sources on scalp-EEG (left) and MEG (right) shown as isocontour maps. The spatial patterns of the N20-P30 source on scalp-EEG and of the SEF_{20}-SEF_{30} source on MEG showed tangential dipolar patterns which were orthogonal to each other (top row). On the contrary, the spatial field patterns of the P25-N35 source on scalp-EEG and of the SEF_{25}-SEF_{35} source on MEG were different. The P25-N35 source showed a radial pattern on scalp-EEG, whereas the SEF_{25}-SEF_{35} source was tangential. Calibration: 10% isocontour lines; maximum of each map = 1 µV on scalp-EEG and 100 fT on MEG; polarities on scalp-EEG: positive (+) solid, negative (–) dashed; field directions on MEG: magnetic field exiting from the head: solid, magnetic field re-entering the head: dashed

focal as the distance between the magnetic field extrema was 5.1 cm and the distance between the electric field extrema was 8.9 cm. On the contrary, the spatial field patterns of the P25-N35 source on scalp-EEG and of the SEF_{25}-SEF_{35} source on MEG were different. The P25-N35 source showed a radial pattern on scalp-EEG, whereas the SEF_{25}-SEF_{35} source was tangential. It should be noted, that is impossible that the SEF_{25}-SEF_{35} could have shown a radial pattern as only its tangential components were measured on MEG. Furthermore, the variance contribution of the SEF_{25}-SEF_{35} source to the evoked response was much smaller than that of the SEF_{20}-SEF_{30} source (68.5% for the SEF_{20}-SEF_{30} source; 16.4 % for the SEF_{25}-SEF_{35} source). On scalp-EEG, the variance contributions were distributed more equally between the two sources (42.7 % for the N20-P30 source and 49.4 % for the P25-N35 source). It should be noted that the variance contributions of the individual sources do not simply sum up to the variance explained by both sources together as the sources are not necessarily orthogonal to each other.

Concerning source locations, scalp-EEG and MEG localization estimates showed close agreement. All sources were within 16 mm of the C4

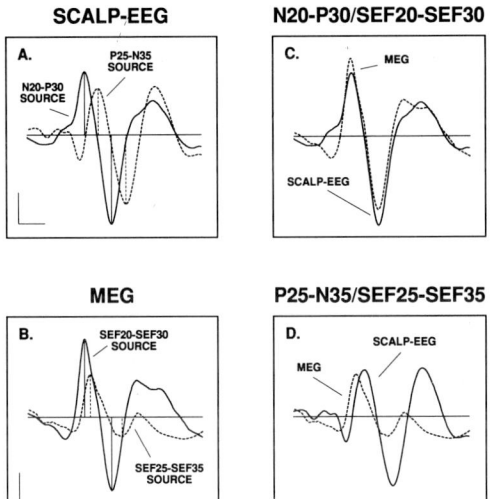

Fig. 4.2.4.6. A., B. Superimposed time patterns of the N20-P30 and the P25-N35 source on scalp-EEG (A), and of the SEF_{20}-SEF_{30} and the SEF_{25}-SEF_{35} source on MEG (B). Vertical lines indicate peak latencies. Calibration: horizontal = 10 msec, time scale begins at stimulus onset; vertical = 1 μV (scalp-EEG) and 50 fT (MEG). **C., D.** Superposition of the time activities obtained with the different techniques for the N20-P30 and the SEF_{20}-SEF_{30} source (C), and for the P25-N35 and the SEF_{25}-SEF_{35} source (D). The time patterns of the electric N20-P30 source and the magnetic SEF_{20}-SEF_{30} source showed almost identical time activities (C), whereas time activities of the electric P25-N35 source and the magnetic SEF_{25}-SEF_{35} source were different (D)

position of the International 10-20 System. The distance between the two sources was 8 mm on scalp-EEG, and 14 mm on MEG. Additionally, depth estimates showed similar features. Both the electric N20-P30 and the magnetic SEF_{20}-SEF_{30} source were located deeper compared to the electric P25-N35 and the magnetic SEF_{25}-SEF_{35} source. The following absolute depth estimates below the cortical surface were obtained: N20-P30 source: 7 mm; SEF_{20}-SEF_{30} source: 9 mm; P25-N35 source: 4 mm; SEF_{25}-SEF_{35} source: 3 mm.

Concerning time activities, the electric N20-P30 source and the magnetic SEF_{20}-SEF_{30} source showed almost identical time activities (Fig. 4.2.4.6C), whereas time activities of the electric P25-N35 source and the magnetic SEF_{25}-SEF_{35} source were different (Fig. 4.2.4.6D).

4.3. Discussion

4.3.1. Cortical Stimulations – Median Nerve Somatosensory Evoked Potentials on Electrocorticography

Our results suggest that cortical stimulations and SEPs recorded on ECoG in conjunction with dipole modeling are useful to study functional anatomy of human somatosensory cortex and provide objective criteria for localization of central fissure in patients undergoing surgery adjacent to sensorimotor cortex.

4.3.1.1. Cortical Stimulations

Cortical stimulations yielded motor responses anterior to central fissure and sensory responses posterior to central fissure. However, stimulation of some electrodes yielded ambiguous results as stimulation anterior to central fissure yielded sensory responses and stimulation posterior to central fissure motor responses. Some electrodes were classified as mixed motor and sensory. The findings of motor responses in postcentral gyrus and sensory responses in precentral gyrus agree with the classical study of Penfield and Boldrey [278]. In their landmark study, these authors concluded that 'movement had a proportionally larger representation anterior to central fissure, and sensation a larger representation posteriorly, and that the two areas, motor and sensory, overlap each other consistently and correspond to each other horizontally'. Later, Penfield and Rasmussen [280] demonstrated that these results were not due to spread as similar results – namely sensory responses from motor cortex – could be obtained in a patient after excision of postcentral gyrus. Animal studies documented a direct thalamic input to primary motor cortex [16, 17]. Furthermore, Goldring and Ratcheson [148] performed single cell recordings in the hand area of human motor cortex and found that cells were activated by active and passive hand movements. These findings document that somatosensory cortex does not exclusively receive sensory information and that motor cortex is not solely providing motor output, but that there exists a sensorimotor region adjacent to central sulcus with complex interactions. In this region, the postcentral gyrus is predominantly sensory and therefore has been termed Sm I (sensorimotor) whereas the precentral gyrus is predominantly motor and therefore has been called Ms I (motor-sensory) according to Woolsey [380].

Although primary motor and somatosensory cortex organization is usually assumed to be quite uniform, the reports in the literature concerning its spatial and specifically anterior-posterior extension are surprisingly variable. In Penfield and Boldrey's classical studies [278], motor and sensory responses were restricted to rather small strips anterior and posterior to central sulcus, respectively. In agreement with these results, Gregorie and Goldring [149] suggested from findings in patients undergoing selective excisions adjacent to central fissure, that the posterior portion of precentral gyrus corresponding to area 4 played a dominant role in producing movement. This is line with experiments in monkeys, where the lowest stimulus intensities for eliciting motor responses were found in the region bordering central sulcus [211]. Furthermore, area 3b in the posterior wall of central sulcus seems to be most important for intact sensory perception in patients with focal excisional surgery [149] and for the learning of somatosensory discrimination [298]. On the contrary, Lüders et al. [228] found that the sensorimotor cortex was an extensive area occupying up to 5-6 cm in the anterior-posterior direction. Furthermore, these authors reported that all the cortex in an extensive perirolandic region was of functional significance and that there was no 'silent cortex' between essential cortical areas such as motor cortex or Broca's area. Our results were somehow in between as we could obtain primary motor and sensory responses approximately from a region extending 3-5 cm in the anterior-posterior direction. In agreement with Lüders et al. [228], we also failed to identify 'silent' cortical areas at the lateral frontal and parietal neocortex.

Furthermore, there is intersubject variability of absolute location of sensorimotor cortex. Already Penfield and Boldrey [278] noted that 'if a point on the precentral gyrus two-thirds of the distance from the longitudinal fissure to the fissure of Sylvius be stimulated in 6 different individuals the results will vary greatly so that any standardized chart which localizes the position of those points upon the cortex may be correct for one individual but not for all'. Ojemann [261] confirmed these findings. He observed variability in the location of sensorimotor cortex in several aspects. First, he found a large interindividual variability in the anterior-posterior location of sensory-motor responses in relation to anatomical landmarks, i.e. from 3 cm posterior to pterion measured along the sylvian fissure to 9 cm posterior to that landmark, in one case with a presumably congenital cortical abnormality at the end of sylvian fissure. Second, the distance from sylvian fissure to the tongue motor and sensory areas was variable across patients. The number of patients in our study was to small to comment on these anatomical variabilities.

Concerning the side of response, Penfield and Boldrey [278] reported no isolated ipsilateral sensations or movements. Bilateral sensations occurred in the face and in the tongue on rare occasions. Ojemann [261] observed ipsilateral face and tongue responses in a considerable proportion of patients. We observed strictly contralateral motor and sensory responses. Occasionally ipsilateral face sensations were elicited probably due to direct stimulation of trigeminal nerve or of the dura. This findings and interpretations are similar to those proposed by Lüders et al. [228].

Somatosensory experiences elicited by cortical stimulation were described by our patients as numbness or tingling most often. This is in agreement with the results of Penfield and Boldrey [278]. These authors observed the quality of cortical sensation as follows. In 204 responses it was called tingling or electricity, in 131 numbness, in 49 instances the patients had a sense of movement when no objective change of position occurred, and finally 11 times pain and 13 times cold were reported. Ojemann [261] reported that his patients described predominately tingling or shock-like sensations. The common occurrence of paresthesias can be explained by the fact that cortical stimulations result in a non-physiological depolarization of cortical cells which differs from the well-organized spatiotemporal activation patterns during physiological stimuli [228].

4.3.1.2. Data

SEPs showed P20-N30 waveforms anterior, and N20-P30 waveforms posterior to central fissure. Both components (P20-N20, termed N20 in the following, and P30-N30 termed P30 in the following) showed a phase reversal across central fissure and primarily tangential dipolar patterns. Furthermore, positive P25 components and negative N35 components were recordable. These components did not show a clear phase reversal and were oriented more radial. There was some intersubject variability concerning waveforms which was probably due to age differences and differences in lesion type, size and location (Figs. 4.2.1.2D and 4.2.1.4D). These differences are similar to other studies which showed intersubject variability in evoked responses on ECoG [145, 146, 149, 378].

In the following our data will be viewed shortly in the light of previous SEP studies on ECoG. It should be mentioned, however, that the findings of some of these studies are not readily comparable to our results due to different recording techniques. Woolsey et al. [382, 384] and Jasper [179] were the first to report recording of SEPs from the cortical surface of human somatosensory cortex using mechanical stimulation. SEPs on human cortex following percutaneous peripheral nerve stimulation were described by Jasper et al. [181] using single sweeps of a cathode ray oscilloscope. Since then, the techniques of recording were developed from averages or summations of many trials with various computing devices usually with limited number of channels [114, 115, 139, 145, 172, 204, 335] to the use of sub- or epidural grid electrodes which allowed a more complete coverage of somatosensory and adjacent cortical areas. These grids have been used either intraoperatively [5, 8, 146, 149, 378] or were implanted chronically thus avoiding time constraints in the operating room [32, 152, 218, 226, 228, 231, 346].

Generally, it is agreed that activity around 20 msec represents the arrival of the thalamocortical volley at the cortex [7, 8, 32, 226, 231, 275, 346, 378]. A phase reversal for this activity with postcentral N20 and precentral P20 waveforms was first observed by Broughton et al. [49, 50, 51]. These findings of a polarity inversion for the N20 component agree with our results and were subsequently replicated by other investigators [4, 5, 7, 8, 11, 32, 226, 231, 275, 346, 378]. Therefore, several authors attributed the N20 component to activity in the posterior wall of central sulcus corresponding to area 3b [4, 5, 7, 8, 11, 32, 51, 226, 231, 346, 378]. In some studies it was noted that the precentral P20 component peaked slightly later than the postcentral N20 component. Some authors explained this phenomenon by volume conduction effects, overlapping activity from distant generators, or by spatiotemporal activation patterns in local neuronal networks [4, 7, 8, 51]. Other investigators suggested additional activity in precentral gyrus [275, 276].

Furthermore, our findings of postcentral P30 and precentral N30 waveforms with phase reversals across central sulcus are in agreement with previous ECoG studies [4, 5, 7, 8, 11, 346, 378].

Similar to our results, positive P25 components were found by other investigators. In a large series of 46 patients, the Yale group around Allison and Wood [5, 8, 378] found P25 components with peak latencies in the 20-30 msec latency range. These components were variable both within and between patients. Whereas in some patients P25 appeared as an additional intermediate peak on an otherwise typical N20-P30 waveform (35% of the cases), in other patients it was considerable larger than N20-P30 and was thus the dominant waveform (30% of the cases). P25 always seemed to be largest at postcentral locations somewhat medial to the largest 20 and 30 msec potentials. These findings of a more medial location of the P25 component compared to the N20-P30 was replicated by Sutherling et al. [346]. In agreement with these results, the three-dimensional localization estimates of the P25-N35 source were medial to those of the N20-P30 in our study. The Yale group proposed that the P25 component was generated in the anterior crown of postcentral gyrus corresponding to area 1 [4, 5, 7, 8, 378]. Furthermore, Allison et al. [8] differentiated between a medial P25 component and a more lateral P25-like component and suggested sequential medial to lateral activation of spatially distinct regions in area 1. This hypothesis is compatible with findings in area 1 of monkey somatosensory cortex where two and sometimes three separate patches of 2-deoxyglucose labeling are evident following flutter stimulation of a single finger [191]. There is also evidence of serial activation of these patches by corticocortical pathways within area 1 [190]. Another finding in the studies of the Yale group [8, 378] was similar to our results: The P25 components in these studies exhibited a significant tangential component besides their predominantly radial activity in some patients. Lüders et al. [226, 231] reported a prominent P2 component at latencies of approximately 24 msec. This P2 component thus can be regarded as the equivalent of the P25 activity. This component showed a larger amplitude and a more rapid fall off compared to the N1 component (equivalent of the N20) suggesting activity in the surface cortex. Furthermore, the P2 component ('postrolandic P2') showed a phase reversal across central fissure. However, the reversed precentral pole ('prerolandic reversed P2') was of significantly lower amplitude, with a relatively more extensive field of distribution and with a gradual fall off over space. In contrast to the Yale group, these authors therefore suggested multiple pre- and postcentral generators for the P2 component. These issues will be further discussed in Section 4.3.1.5.

In contrast to our findings and the results of the aforementioned studies [4, 5, 7, 8, 32, 51, 226, 231, 275, 346, 378], other groups could not observe a polarity inversion for the first cortical activity between pre- and postcentral recording sites, but instead reported on a P1 component with a single electropositivity [59, 204, 335]. This discrepancy could possibly be explained as follows. First, these authors reported an initial positive deflection with peak latencies around of 25-30 msec [335] and 28.4 msec [59]. Second, sometimes smaller preceding negative peaks were observed which persisted after superficial cortical excisions [335]. Third, Lüders et al. [226, 231] noted that the N1 amplitude (N20 equivalent) was significantly smaller than the P1 amplitude (P25 equivalent). Therefore, it could be possible that the P1 component of these authors actually is equivalent to the P25 component and that the N20 component simply was missed in some patients due to its comparably smaller amplitude or incomplete exposure of somatosensory cortex during surgery.

In the present study, we performed referential recordings from the cortical surface against an uninvolved reference. On the contrary, the group around Goldring used bipolar recordings and observed three types of responses [146, 149]. First, in some patients responses could be recorded from somatosensory cortex exclusively. In these cases, SEPs were seen only on two adjacent pairs of electrode combinations, and these responses were reversed in polarity. Second, in other patients responses could be obtained from both somatosensory and motor cortex. This resulted in responses from 4 adjacent electrode pairs with a double phase reversal. Finally, in a small number of cases a phase reversal was not seen and only responses of varying amplitude from several adjacent electrode combinations occurred. In these instances, the largest amplitude response usually was seen in the somatosensory area. Which kind of response could be recorded was independent of anesthesia and the side of recording – whether dominant or non-dominant hemisphere [146, 149]. Phase reversals in bipolar recordings are equivalent to amplitude maxima in referential recordings. In agreement with other authors [8, 226, 231, 378], we always could obtain precentral potentials. This discrepancy to the studies of Goldring and Gregorie [146, 149] could result from placement of precentral electrodes lateral to the P20-N30 maximum, especially because bipolar electrode chains were oriented parallel to the midline. Recently, King and Schell [206] reported on difficulties in identifying pre- and postcentral gyrus using bipolar montages. Thus, in agreement with the recommendation of Wood et al. [378] we think that referential recordings are superior to bipolar recordings for exact delineation of central fissure.

Another method to study SEPs invasively has been transcortical recording from the cortical surface and the underlying white matter [145, 204, 335]. From postcentral sites, surface positive waves with a polarity inversion in the white matter could be recorded which is indicative of generation of this component in postcentral surface cortex. The peak latencies of these waves were in the range of 18-25 msec [204] and of 25-30 msec [335] which fits to the latencies of the P25 component observed in the present study. In the initial study performed by Kelly et al. [204], transcortical recordings yielded no responses from precentral sites. However, in subsequent studies from the same group [145, 335] polarity inversions between the cortical surface and the underlying white matter were found in some patients. These potentials either had the same latency compared to the postcentral waveforms, or were delayed for about 2-4 msec and then showed no perfect phase reversal between surface cortex and white matter [145]. These findings point towards a local generation of this component in the anterior crown of precentral gyrus. However, it should be noted that ablation of somatosensory cortex led to a complete loss of median nerve SEPs which contradicts a significant precentral generator [8, 9, 10, 335].

4.3.1.3. Source Localization Techniques

An ultimate understanding of SEPs requires knowledge about their underlying neuronal sources [7]. Previous investigators have proposed different theories concerning the neurogenesis of early SEPs which yielded partially converging evidence and partially led to still unresolved controversies. Most of these investigators based their conclusions concerning the neuronal sources on classical EEG criteria like phase reversals, maximum amplitude, or waveshape thus applying a somehow phenomenological approach [4, 5, 51, 59, 145, 146, 149, 204, 226, 231, 275, 276, 335, 378]. However, electrical fields in the brain obey complicated biophysical laws [258] and thus this approach may be inaccurate. On the contrary, dipole modeling techniques take into account these physical laws and thus should provide accurate three-dimensional intracerebral localization estimates of the neuronal sources underlying SEPs. In these models, explicit assumptions concerning the volume conductor and the configuration of the source have to be made (for further details see Section 2.2). Thus, these methods should be useful to provide objective hypotheses concerning the neurogenesis of the somatosensory evoked response. Sutherling et al. [346] conducted to our knowledge the only study where dipole modeling was applied to SEP data on ECoG. These authors modeled the N20, P25, and P30 component and achieved excellent localization estimates of central fissure. However, a shortcoming in their study was the use of single equivalent dipole models. The basic modeling assumption of this approach is that only one brain area is active at the time points which are used for modeling. However, most authors agree that the human somatosensory evoked response is generated by multiple sources which overlap both in space and time [7, 8, 32, 226, 231, 275, 346, 378]. Nunez [259] has shown that modeling activity generated by multiple sources with a single dipole can lead to erroneous results. Indeed, the single dipole model in the study of Sutherling et al. [346] often yielded inadequate fits. Furthermore, these authors restricted their analysis to single time slices give by peak amplitudes. We therefore extended these previous studies and applied a spatiotemporal model which allowed to study multiple simultaneously active sources over the entire time domain of the evoked response.

4.3.1.4. Spatiotemporal Modeling

Our findings using spatiotemporal modeling suggest that the majority of human SEPs on ECoG are generated by two dipolar sources in the postcentral gyrus which overlap both in space and time. Multiple source modeling was useful to determine the number of sources, their locations, their time activities, and their spatiotemporal field patterns. The results were similar across patients. Reproducibility of data within patients, with low run-to-run variability, further supports the biological reliability of our results. We considered two sources as sufficient for several reasons. First, the two source model reproduced the data very closely explaining an average of

84.6 % of the variance. Second, adding a third source did not significantly increase the variance accounted for above noise. Finally, principal component analysis revealed two significant covarying processes underlying the data. The localization estimates were supported by convergent evidence from cortical stimulations, intraoperative photographs, and normal sensorimotor examination after focal excisional surgery. Specifically, there was good agreement with the results of cortical stimulations as the surface projections of the source locations were close to electrodes which elicited sensory experiences in the hand. The time activities of both sources showed considerable overlap. We found that one source generated the N20-P30 component, and the other generated the P25-N35 component which is in accordance with the results on scalp-EEG and MEG in this study (Sections 4.2.2.3 and 4.2.3.3).

Furthermore, multiple source modeling was useful to separate the spatiotemporal potential patterns of the two sources, thus revealed an underlying simplicity of the evoked response, and gave quantitative estimates of the total number of active sources. Thus, spatiotemporal modeling can be regarded as a useful tool to form testable hypothesis about somatosensory information processing in the human brain both in space and time.

4.3.1.5. The Neuronal Sources Underlying SEPs

We suggest two generators of somatosensory evoked potentials, one horizontal dipole in the posterior wall of the central fissure (area 3b), generating the N20-P30 component, and one radial dipole in the anterior crown of the postcentral gyrus (area 1), generating the P25-N35 component. The N20-P30 source was located in postcentral gyrus and was oriented primarily tangentially with a phase reversal approximately following central sulcus in all patients. The mean depth of this source was 8 mm below the cortical surface. Allison et al. [8] obtained a mean depth of 18 mm for the central sulcus from 6 human brains. Thus, our results are compatible with activity at the centroid of the posterior wall of central sulcus. Spatiotemporal modeling suggests identical or at least similar neuronal populations generating the N20 and P30 components. This is compatible with the hypothesis that the N20 component is generated by the initial depolarization of the pyramidal cell bodies and the proximal apical dendrites, whereas the P30 component is generated by the later depolarization of the distal portion of the apical dendrites [78, 212, 365, 373].

The P25-N35 source also was located in postcentral gyrus in all patients. However, the orientation of the P25-N35 source was variable across patients and showed significant tangential components in some patients (cf. Patient #1), whereas in other patients predominantly radial components were seen (cf. Patient #2). This finding is similar to ECoG data in some patients of other studies [7, 378], but different from the ECoG data of Sutherling et al. [346] who found predominantly radial patterns for the P25 component in all patients. The tangential components of the P25-N35 source could be explained by the fact that area 1 is located in the anterior half of postcentral gyrus, but also occupies upper parts of the posterior wall of central sulcus [293, 361]. Furthermore, we found an average depth estimate of 8 mm for the P25-N35 source, which is identical to that of the N20-P30 source. Such a depth estimate is not compatible with activity in surface cortex. Furthermore, these depth estimates disagree with the scalp-EEG and MEG results in this study where the P25-N35 source was localized more superficially in comparison to the N20-P30 source. Furthermore, previous studies on ECoG using single moving dipoles to model selected time points found deeper location estimates for the N20 and P30 sources, and more superficial depth estimates for the P25 and N35 sources [346]. These differences could be explained as follows. First, the configuration of the two sources could change over the time series, due to a center of mass averaged over time. Second, modeling an extended source by an equivalent dipole results in a deeper depth estimate [263]. Finally, ECoG measurements are made rather close to the source which could violate some basic model assumptions of the dipole concept, i.e. that measurements are made at a distance which is large compared to the dimensions of the source [258]. This holds especially true for superficial cortical activity like in area 1. In conclusion, the causes for these differences warrant further investigation. Like for the N20-P30 source, we propose identical or at least similar neuronal generators for P25 and N35 activity reflecting sequential activation of the proximal and distal apical dendrites of cortical pyramidal cells [78, 212, 365, 373].

Our results are in accordance with those of other investigators. The hypothesis of a tangential dipole in area 3b generating the N20 and the P30 components was originally proposed by Broughton et al. [49, 50, 51] based on a combined scalp-EEG and ECoG study. Subsequently, Allison et al. [7] agreed with this hypothesis and additionally suggested a radial source in area 1 generating the P25 and the N35 components. These hypotheses were subsequently confirmed by further results from the Yale group [4, 5, 7, 8, 378] and the UCLA group [32, 346]. Whereas the neurons in area 3b are arranged in the anterior posterior direction which fits to a tangential dipole, the neurons in area 1 are primarily arranged perpendicular to the cortical surface compatible with a radial dipole [293, 361]. Concerning afferent inputs from the ventral posterior lateral nucleus of the thalamus (VPL), anatomical studies using the Nauta technique have revealed that area 3b receives a very dense projection of large diameter fibers, whereas area 1 receives a sparse projection of smaller diameter fibers [185, 188, 294]. This would explain the latency shift between the N20-P30 source and the P25-N35 source which is also in line with the results of animal experiments [10, 14, 391].

The hypothesis of N20 as generated in area 3b agrees with the interpretation of Lüders et al. [226, 231], who referred to this component as N1. N20 usually has a more widespread field pattern with a more gradual fall off over space which suggests a source in the sulcus rather than at the cortical surface. Furthermore, area 3b is the major recipient of thalamocortical afferents [67]. Kelly et al. [204] using transcortical recordings from postcentral gyrus found that N20 showed no phase reversal between cortical surface cortex and the underlying white matter and that this component persisted after focal excision of the cortex immediately underlying the recording electrode. Topical application of procaine or veratrine hydrochloride abolished the P2 component (equivalent to N20), but did not affect the N1 component (equivalent to P25) [147]. All these findings strongly support the neurogenesis of N20 in area 3b. On the contrary, Papakostopoulos et al. [275, 276] suggested that N20-P30 and P20-N30 are generated by two independent radial dipoles in somatosensory and motor cortex. They found that pre- and postcentral waveforms were differently affected by centrifugal activity, i.e. externally-paced (passive) finger displacement, self-paced (voluntary) finger displacement, and electrical median nerve stimulation during self-paced finger displacement [273, 274]. Celesia et al. [59] made similar conclusions based on their SEP recordings from human thalamus and from the cortical surface. These authors attributed the N_0 component (the equivalent of N20) to thalamocortical projections, rather than to cortical activity. In conclusion, at the present time most investigators believe that N20 is the first cortical activity and is generated in area 3b. This interpretation is further supported by scalp-EEG and MEG data from humans [42, 100, 109, 162, 346, 352, 375] as well as by various animal experiments [13, 14]. These issues will be discussed in further detail in Section 4.3.6.

The neurogenesis of the P25 component is still more controversial. The Yale group around Allison proposed a radial generator in area 1 [4, 5, 7, 8, 378] which is in line with our interpretation [32]. Lüders et al. [226, 231] agreed with this hypothesis only partly. As already mentioned (Section 4.3.1.2), these authors suggested that the P2 component (equivalent to the P25 component) was generated in areas 1, 2, 3a, and 4. Activation of areas 1 and 2 and of the crown of area 4 was proposed to generate the major classical primary positivity of the P2 component ('postrolandic P2') seen best in the postcentral gyrus (areas 1 and 2) and the immediate precentral region (crown of area 4). The precentral relatively low-amplitude negative pole of P2 ('prerolandic reversed P2') would be due to the rostral pole of horizontal dipoles in areas 3a and 4 in the anterior bank of central fissure. Transcortical recordings constantly demonstrated phase reversals between the cortical surface and the underlying white matter in postcentral gyrus indicative of activity in area 1 [145, 204, 335]. Furthermore, in some patients similar observations also could be obtained from precentral gyrus pointing towards activity in the crown of area 4 [145, 204, 335]. However, focal excisions of somatosensory surface cortex underlying the recording electrode abolished the P2 component in humans [9, 335], whereas removal of motor cortex had little effect on SEPs in monkeys [10]. In conclusion, we think that the contribution of area 1 to the P25 component seems to be rather undisputed whereas the contribution of other cytoarchitectonic structures remains to be clarified in future studies.

4.3.1.6. Limitations of the Procedure

There also were disadvantages of our approach. One limitation of the analysis used is modeling of activity which occurred from 14 to 17 msec as one of two cortical sources active from 20 to 60 msec. Activity from 14 to 17 msec likely arises from brainstem and thalamocortical radiations prior to the arrival of the sensory volley at the cortex [65, 98]. Inclusion of similar polarity activity at these latencies in both sources in Patient #2 (Fig. 4.2.1.3C and 4.2.1.4B) suggested the presence of activity from sources other than only two cortical dipoles. This activity did not occur in all patients (cf. Patient #1). Similar variability of occurrence of additional early components has been noted by other investigators [7]. Earlier SEP peaks can be enhanced by high pass filtering and attenuated by low pass filtering [233]. We used a bandpass 1-1000 Hz similar to previous investigators [378] which is different from some methods for intraoperative SEPs on ECoG which use a bandpass of 30-3000 Hz and which routinely find small, earlier peaks [260]. The restricted spatial extent of some of these potentials also could result in variability of appearance due to the 1 cm interelectrode distance we used. However, since our question was the number of sources active, we did not exclude activity due to a-priori assumptions and modeled the entire time series. It is difficult to determine more about the nature of the generators of these earlier potentials without more complex models or subcortical electrodes, which were not used in these patients for clinical diagnostic reasons.

4.3.2. Median Nerve Somatosensory Evoked Potentials on Scalp-EEG

4.3.2.1. Data

In our study, SEPs were reproducible within subjects, but showed some variability between subjects. This intersubject variability is similar to our findings on ECoG reported previously [32] and in this study, as well as to the results on scalp-EEG by other authors [94, 100, 142]. Nevertheless, the data showed some consistent features concerning prominent SEP components which were reproducible in all subjects which will be discussed in this section. Furthermore, spatiotemporal modeling yielded highly reproducible and consistent results across subjects which further underlines biological reliability of our data.

Scalp SEPs showed some early activity occurring before the first cortical activity reflected by the N20 component. This is in agreement with the literature. These subcortical potentials consist of three positive far-field potentials (P9, P11, and P14), followed by a negative N18 component [73, 77, 98, 238, 240, 359] and were first recorded by Cracco [73]. Whereas the P9 component reflects the brachial plexus volley [66, 77, 98, 104, 230, 386], P11 is the far-field equivalent of the spinal N11 [12, 13, 97, 98, 104]. The neuronal generators of the P14 component were attributed to the afferent volley in medial leminiscus [12, 66, 103, 104, 105, 108, 135, 230, 237, 238, 239, 240, 387]. The negative N18 component can be recorded in a widespread pattern bilaterally over the scalp. Desmedt and Cheron [105] were the first to distinguish the bilateral N18 from the contralateral N20 cortical response. As evidenced from patients with hemianesthesia and thalamic or suprathalamic lesions, the N18 component is believed to be generated subcortically [240]. Mauguière and Desmedt [238] studied patients with long-standing hemispherectomy without damage of the striatum or the diencephalon which precludes cortical evoked potentials. Stimulation of the side opposite to the missing hemisphere evoked a slow N18 negative potential of 15 to 25 msec duration bilaterally. Long-standing hemispherectomy entails massive retrograde degeneration of thalamo-cortical neurons. Therefore, the authors concluded that the N18 component reflects neuronal activity below the thalamus and is volume-conducted to the scalp. Furthermore, animal studies suggest multiple generators in the brainstem, namely the tectum and pretectum, for this component [19, 46, 158]. Recently, Urasaki et al. [359] recorded SEPs from the thalamus and the ventricular system during stereotaxic operations in humans and postulated that N18 is generated in brain-stem between the upper pons and the mid-brain which is in line with the results of Mauguière and Desmedt [238]. Due to its widespread pattern over the entire scalp, the N18 component can be best identified when scalp-EEG is referenced to an uninvolved non-cephalic reference [103, 105]. As our study was focused on cortical SEP components, we used linked earlobes as reference which reduces volume conducted far-field potentials as P9, P11, P14 and N18 components. Therefore, we could record only minor subcortical activity and did not analyze these components (cf. Subject #2)

The next SEP component which – in contrast to the N18 component – can be recorded only contralaterally with a maximum over the parietal region is the N20 component. It is generally agreed that the N20 component represents the first cortical activity and reliably identifies the arrival of the somatosensory volley at the cortex [8, 32, 42, 94, 100, 109, 346, 375]. We could record N20 components at the parietal electrodes which were accompanied by positive mirror images – the P20 components – at the frontal recording positions. These waveforms resulted in tangential isopotential maps at 20 msec. The phase reversal was slightly oblique from anterior inferior to posterior superior and thus followed approximately the central sulcus which is in line with previous studies [94].

Whereas latency, waveform, and spatial potential distribution of the N20 component is pretty much undisputed in the literature, the later SEP components are more controversial. After the N20 component, we found the following wave sequences. First, we could record a P25 component which showed a single electropositivity with no clear phase reversal and could be best seen at the central (cf. Subject #1) or parietal electrodes (cf. Subject #2). Isopotential maps at these latencies resulted in a primarily radial pattern. It should be mentioned, however, that this pattern was sometimes obscured by overlap from the preceding and following waveforms.

Second, we could identify a positive P30 component which could be most clearly seen over contralateral parietal cortex. Similar to the N20 component, this P30 component was accompanied by a nearly mirror image at the frontal recording positions resulting in a frontal N30 component. Similar to the N20-P20 components, isopotential maps at 30 msec showed tangential dipolar patterns with a phase reversal following central sulcus. N20 and P30 isopotential maps were of reversed polarity. Although the P30 isopotential maps had similar spatial potential distributions compared to those of the N20 component, these maps were not identical. This suggests either that different neuronal sources generate these components, or that activity from multiple brain areas is superimposed to form the P30 component. This issues will be further discussed in Sections 4.3.2.3 and 4.3.2.4.

Finally, we could record a N35 component which was superimposed on the slope of the P30 component and showed a single electronegativity. Isopotential maps at these latencies showed radial patterns and were of opposite polarity compared to the P25 component. Concerning the spatial patterns, the P25 and N35 maps were similar, but not identical, which raises the same issues as for comparison of N20 and P30 maps mentioned above. For the sake of simplicity, we will refer to these components as N20 (comprising frontal P20 and parietal N20), P25, N30 (comprising frontal N30 and parietal P30), and N35 components. Our findings are in agreement with previous studies from our group [346] and are in line with the results of the Yale group [4, 5, 7, 8, 378].

However, our findings disagree with some results of other investigators. Similar to our results, Cracco et al. [73, 74, 75, 76, 77] found N20 components which were most prominent at scalp locations overlying the specific somatosensory cortex contralateral to the side of stimulation. Furthermore, in these studies positive P25 components could be identified. However, in contrast to our results, these authors observed a progressive increase in peak latency for these components from anterior to posterior in the sagittal plane, and from contralateral to ipsilateral to the side of stimulation in the coronal plane. In the sagittal plane these latency differences were most pronounced across the central sulcus. The authors introduced the term 'traveling waves' for these findings, and explained this phenomenon by two or more generators which are activated closely to each other but not simultaneously in time. Similar observations have been made by Yamada et al. [385, 386, 387] who found bifrontal N17-P20-N29, central N19-P23-N32, and parietal N20-P26-N34 components. As these components were sometimes independently affected by lesions, these authors suggested multiple, at least partially independent thalamocortical projection systems. These findings serve as motivation to apply spatiotemporal modeling which can further resolve the question of multiple simultaneously active brain sources overlapping both in space and time. Nevertheless, the traveling wave concept has been left in most recent publications on scalp-SEPs.

Desmedt and colleagues are one of the leading groups concerning somatosensory evoked potential research on scalp-EEG and have contributed numerous substantial publications to the field [97, 98, 99, 100, 101, 102, 103, 104, 105, 106, 107, 108, 109, 110, 237, 238, 239, 240]. In agreement with our results, they found parietal N20 components which were accompanied by frontal P20 components with a phase reversal following central fissure [94, 99, 100]. The later components recorded by this group, however, somehow differ from our results. Desmedt and Cheron [105, 106] identified a focal midfrontal P22 component which appeared to be distinct from the P20-N20 component. The findings of a separate P22 component were further substantiated by different effects of aging [105, 106] and interstimulus intervals [95] on N20 and P22 components, as well as by color imaging techniques [94, 99, 100], and, most important, by the lesion experiments conducted by Mauguière et al. [239] which documented a dissociated loss of N20 and P22. After the P22 component, these authors observed a large bilateral N30 component frontally and a focal P27 in the parietal region [100]. Thereafter, a P45 component appeared over the central scalp in about half of the young subjects, whereas it was recorded more regularly in old normals [100, 104]. When present, P45 always involved the contralateral central region but it could extend either towards the front or back ipsilaterally [100]. In conclusion, these authors propose the following sequence of cortical SEP components: N20, P22, P27, N30, and P45, where N20, P27, and P45 were attributed to parietal structures, and P22 and N30 to frontal structures. The discrepancy to our results could be explained in several ways. First, some of these components could actually reflect activity from multiple simultaneously active brain sources which could lead to a shift both in location and latency [4, 8, 40, 259]. Second, we used different spatial recording strategies compared to these investigators. Desmedt recorded from the whole scalp using 16 [100] to 27 electrodes [109] according to a modified 10-20 System. We recorded from 32-48 electrodes arranged in a rectangular array centered at the C3 or C4 position with 30 mm spacing. Thus, we may have missed contralateral, midline (i.e. activity from the supplementary motor area (SMA)), or widespread activity. However, we were mainly interested in activity generated in primary receiving areas. The spatial resolution of our recording electrodes overlying primary motor and somatosensory cortex was higher than in Desmedt's studies and it is unlikely that we missed local components due to spatial aliasing. In order to study more widespread or bilateral components, however, different electrode placements than those used in our study would have to be used. Finally, there were differences in temporal resolution. We used a bandpass of 1 Hz to 1000 Hz and a sampling frequency of 4 kHz. Desmedt et al. [100, 109] used a bandpass 1.6 Hz to 5 kHz and a sampling frequency of 5 to 8 kHz for their studies which could have resulted in a better temporal resolution. Furthermore, Desmedt et al. [102] recommended a bandpass from 1.6 Hz to 2 kHz and a sampling frequency of 4 kHz for recording of SEPs.

4.3.2.2. Source Localization Techniques

While in the beginning of SEP research only few channels were used for recording, the advent of multichannel systems allowed more detailed analysis of SEPs [100, 120, 142]. Recently, color imaging techniques were introduced into SEP research by Desmedt and colleagues [100, 109].

However, it should be mentioned in this context that brain mapping techniques are a mere graphical display of surface potentials, and do not provide explicit information about the three-dimensional location of the neuronal sources underlying these potentials. These techniques do not dwell into the inverse problem and do not take into account the laws of volume conduction which is a necessary condition to understand processing of somatosensory information in the human brain [141, 258]. Therefore, these maps have to be interpreted with caution. Deiber et al. [94] used a bisection and maximum amplitude technique to study the neuronal sources underlying digit stimulation. Recently, Desmedt et al. [103] compared simulated data from various dipole configurations in a 3-shell spherical model with actual SEP data. However, in this study the inverse problem was not dealt with either. In conclusion, the application of dipole models to investigate the neuronal sources of scalp SEPs has been rather scarce so far. This is surprising, given the widespread use of dipole localization techniques in MEG research and the longstanding availability of solutions for EEG applications [83, 84, 337, 338, 372].

To our knowledge, the only studies where SEPs recorded on scalp-EEG were analyzed with dipole models were performed by Allison et al. [7] and Sutherling et al. [346]. The latter authors compared localization accuracies of scalp-EEG, MEG, and ECoG. In this study, single equivalent dipoles models were used and source localization was restricted to time points given by peak latencies. However, as noted in some theoretical studies, modeling of activity generated by multiple brain sources with a single source can introduce significant errors in localization [40, 259]. As it is generally agreed that early SEPs are generated by multiple neuronal sources overlapping both in space and time, we think that multiple dipole modeling is more adequate to study this problem than single dipole models are.

4.3.2.3. Spatiotemporal Modeling

Our results indicate that short latency somatosensory evoked potentials (SEPs) on scalp-EEG during median nerve stimulation are generated by two neuronal sources producing potential patterns which overlap both in space and time. One source generated the N20-P30 component, the other the P25-N35 component. Multiple source modeling allowed to infer the number of sources, their three-dimensional intracerebral locations, their time activities, and their spatiotemporal potential patterns. Furthermore, it should be stressed that multiple source modeling allowed to separate the spatiotemporal potential patterns of the N20-P30 source and of the P25-N35 source which showed considerable overlap.

Concerning the number of sources, we found that two sources were sufficient to model early median SEPs for several reasons. First, adding a third source did not increase the variance accounted for by the model above run-to-run variability. Second, PCA indicated only two significant covarying parts underlying the system variance. Finally, the model could reproduce the data very closely, accounting for an average of 87.7 % of the data variance.

Concerning source location, we found that both sources were located close to the C3 or C4 position of the International 10-20 System, which is reasonable as this position lies within 15 mm of central sulcus [173, 174, 180, 331]. We thus conclude that multiple source modeling gave good non-invasive localization estimates of hand somatosensory cortex. The average distance between the two sources was 12 mm, which indicates an extended area of activation during peripheral nerve stimulation. This finding is compatible with human studies using SEPs on ECoG [8, 378] and with animal experiments using single cell recordings and 2-deoxyglucose labeling [185, 190, 191].

Concerning time activities, one source generated the N20-P30 components, and the other the P25-N35 components. The time activities of both sources showed considerable overlap. The sources generating the N20-P30 components showed a tangential spatial pattern reversing polarity between the frontal and parietal electrodes. The sources generating the P25-N35 components showed a more radial spatial potential pattern, though in some subjects this source exhibited a considerable tangential component (cf. Subject #2).

4.3.2.4. Neuronal Sources Underlying SEPs

Our working hypothesis is that two postcentral sources explain the majority of SEPs recorded on the scalp. The N20-P30 source is represented by a primarily tangential dipole in the posterior wall of central fissure corresponding to Brodman's area 3b, whereas the P25-N35 source is represented by a primarily radial source in the anterior crown of postcentral gyrus corresponding to Brodman's area 1.

The N20-P30 source showed a tangential dipolar pattern with a phase reversal between frontal and parietal recording electrodes. The phase reversal approximately followed central sulcus. Concerning depth the N20-P30 source was located deeper compared to the P25-N35 source at an average depth of 10 mm below the cortical surface. As already pointed out in Section 4.3.1.5, this depth estimate is compatible with activity at the centroid of central sulcus [8]. These results from spatiotemporal modeling of non-invasive scalp-EEG data confirm our ECoG results and suggest similar neuronal populations generating the N20-P30 sequence reflecting sequential depolarization of proximal and distal apical dendrites of cortical pyramidal cells [78, 212, 365, 373].

The P25-N35 source showed a primarily radial pattern, although in some subjects – as already mentioned – we found a considerable tangential component (cf. Subject #2). This source was located more superficially at an average depth of 4 mm. This depth estimate probably seems too deep for superficial cortical activity. However, as already outlined in Section 4.3.1.5, area 1 lies primarily in the anterior portion of postcentral surface cortex, but also occupies parts of the upper portion of the posterior wall of central sulcus which could explain the tilt from radial and could also partly account for the obtained depth estimates [293, 361]. Nevertheless, depth estimates of 4 mm are still too deep and are probably due to inaccuracies of the simple spherical dipole model. The findings of a considerable tangential component for the P25-N35 source are compatible with our previous results on spatiotemporal modeling on ECoG and MEG [32, 34, 42], and the ECoG and MEG results in this study (Sections 4.2.1.4 and 4.2.2.3).

Our results are in accordance with those of other investigators. As already mentioned in Section 4.3.1.5, the hypothesis of a tangential dipole in area 3b generating the N20 and the P30 component, and a radial dipole in area 1 generating the P25 and the N35 component was proposed and substantiated by the Yale group [4, 5, 7, 8, 11, 51, 378]. Neuromagnetic studies provided further supportive evidence to this hypothesis [34, 42, 346, 375]. For further details on the results of these studies the reader is referred to Sections 4.3.1.2, 4.3.1.5, 4.3.3.1, and 4.3.3.3.

Generally, it is agreed that the N20 component is generated by a tangential source in area 3b whereas the neurogenesis of the later components is more controversial [94, 99, 100, 109, 110, 310]. Neurogenesis of the N20 component in area 3b has been suggested by converging evidence from mapping studies, dipole modeling, and lesion experiments [94, 100, 109, 239, 346]. The only exception has been a recent report by Hashimoto et al. [168] who proposed a radial, rather than a tangential dipole in postcentral gyrus generating the N20 component. In patients with frontal lesions, these authors found a negative potential at the midposition between the frontal and central areas which was synchronous to the parietal N20 component, the so-called NFC-potential. These results are especially in contrast to the classical experiments by Mauguière et al. [239], who could not identify such a NFC-potential in patients with selective frontal lesions.

In contrast to our findings, the group around Desmedt and Mauguière proposed a different neurogenesis for SEP components after the N20 component as follows. First, a precentral P22 component was proposed to be generated by a radial dipole in primary motor cortex corresponding to area 4 [94, 100, 105, 109]. Second, a parietal P27 component was thought to be generated by a radial dipole in area 1 [100, 110]. Third, a widespread frontal N30 component was proposed to be generated in the supplementary motor area (SMA) [100, 109, 110]. Finally, no definitive conclusions were made concerning the neurogenesis of the P45 component [100, 109, 110]. The reasoning for these hypotheses will be presented shortly in the following.

A precentral radial generator for the P22 component was initially proposed by Desmedt and Cheron [105]. These authors found that the precentral N22 component was distinct from the frontal P20 mirror image of the N20 component due to significant differences in latencies. Furthermore, the effects of aging on the N20 and P22 component were different [105]. Using color imaging techniques, Desmedt and Bourguet [100] showed concentric positive isopotentials with no adjacent negativity for the P22 component and therefore concluded that P22 was generated by a radial dipole in area 4. Deiber et al. [94] recorded digit SEPs and found that the P22 component, but not the N20 component shifted more medially over the scalp when the stimulus was moved from contralateral thumb to little finger. Concerning the effect of interstimulus intervals, Delberghe et al. [95] observed that during an increase in stimulus frequency the P22, P27, and N30 amplitudes decreased significantly while N20 remained unchanged. Low concentrations of isoflurane caused a dramatic increase of the P22 amplitude without affecting the N20 component [257]. When stimuli to digits were accompanied by voluntary movements, N20 was not affected whereas P22 and N30 were markedly attenuated [61, 71]. Mauguière et al. [239] studied selected patients with single circumscribed hemispheric lesions and differentiated parietal N20-P27-P45 components from prerolandic P22-N30 components. Specifically, complete parietal lesions produced contralateral hemianesthesia without upper motor neuron signs and eliminated the parietal N20-P27-P45 while the prerolandic P22-N30 persisted at usual latencies. On the contrary, precentral lesions with severe hemiplegia eliminated the prerolandic P22-N30 but did not alter the parietal N20-P27-P45 components. These findings were confirmed by other authors [329]. Such double dissociation through focal lesions suggests distinct neuronal sources for N20 and P22 components. According to Desmedt et al. [100, 109], these findings would also suggest separate thalamocortical inputs to the proposed generator sources of N20 in area 3b and of P22 in area 4. Indeed, there is anatomical evidence in monkey that exteroceptive afferents relay in the ventral posterior lateral nucleus of the thalamus with fast direct projections to area 4 [182, 357]. Additionally, motor cortex neurons respond with short latency to skin and joint afferents from contralateral hand [17, 217, 351]. As already mentioned, for P27 a generator in area 1 is postulated, whereas N30 is thought to reflect activity in the SMA [100, 110]. Recent experiments have shown different susceptibility of P27 and N30 to habituation and selective attention [110, 354]. Whereas area 1 receives direct thalamic input [188], SMA receives cortico-cortical projections from motor areas 4 and from parietal areas 1, 2, and 5, but no short latency exteroceptive inputs from thalamus [182, 184].

Possible reasons for these different interpretations concerning the raw data have already been discussed in Section 4.3.2.1. The different conclusions on the neurogenesis of the evoked response could be explained as follows. First, as indicated from our results on spatiotemporal modeling, scalp-SEPs following median nerve stimulation are generated by multiple simultaneously active brain regions which overlap both in space and time. Thus, conclusions based on the assumption of single, sequentially activated neuronal populations may be misleading as theoretical studies have shown [40, 259]. Furthermore, Allison et al. [8] have shown in a recent study a progressive increase in latency from precentral to postcentral recording positions due to superposition of the primary positivity of area 3b (P20), the primary positivity of area 1 (P25), and the primary negativity of area 3b (P30). These authors concluded that P20 contributed most of the voltage at precentral locations, P25 contributed most of the voltage near central sulcus, and P30 contributed most of the voltage at the postcentral locations. Furthermore, these results of sequential and overlapping activation of areas

3b and 1 could also be observed in animal experiments [10, 13, 14]. Second, Desmedt and coworkers [99, 100, 103] based their conclusions on inspection of sequential isopotential map features. However, they did not apply source localization techniques to actually infer the three-dimensional location and time activities of the neuronal sources underlying SEPs which should be more accurate to draw conclusions concerning the neurogenesis of the evoked response. Third, the hypothesis of a significant precentral P22 source as proposed by Desmedt et al. [99, 100, 103, 106] contradicts the results of ablation studies where surgical removal of hand somatosensory cortex abolished all median nerve SEPs in humans [9, 335] and monkeys [10], whereas removal of hand motor cortex had little effect on SEPs in humans [8] and monkeys [10]. Finally, in Desmedt's model [100, 103, 109] no tangential sources are included in the latency range around 30 msec. This is in contradiction to the results of neuromagnetic recordings. As already mentioned, MEG is sensitive primarily to tangential activity, whereas radial activity ideally produces no detectable MEG signal outside a spherical volume conductor [150, 369, 370]. However, in several MEG studies prominent activity at 30 msec arising from a tangential dipole could be recorded and was termed P27m [352, 353], P30 [34, 42, 346, 375], and SEF_{30} (this study), respectively. These issues concerning the neurogenesis of the human somatosensory evoked response will be further discussed in Section 4.3.6.

To summarize, we think that from scalp-EEG studies alone no definite conclusions concerning the exact anatomical source locations can be made. However, scalp-EEG in conjunction with other techniques like ECoG and MEG can be used to form testable hypotheses concerning these questions.

4.3.2.5. Limitations of the Procedure

Our results should be viewed within the limitations of the procedure. In some subjects, both sources exhibited activity occurring as early as 15 msec corresponding to subcortical activity. As already mentioned in Section 4.3.2.1, there are two possible explanations for the inconsistent and rather small occurrence of subcortical components in our study. First, we used a linked-earlobe reference which is sensitive selectively to local cortical activity, whereas widespread far-field potentials tend to cancel out [103, 355]. Thus, for inferences on subcortical sources a non-cephalic reference electrode should be chosen. Second, a different placement of the recording electrodes covering the whole head and not just a restricted area overlying somatosensory cortex – as in the present study – might be useful to model subcortical components. However, as the main interest of our study was the localization of the cortical sources underlying SEPs on scalp-EEG, we did not further pursue this question.

Other limitations of the modeling procedure might be assumptions concerning the source configuration, i.e. the dipole concept, assumptions concerning the volume conductor, i.e. a concentric 4-shell spherical model, and finally the assumption of spatially fixed sources varying in activity over time. These questions will be addressed in more detail in Section 4.3.5. Here it shall suffice to say that theoretical, experimental, and human studies have shown that these assumptions introduce no major error in localization of the sources in somatosensory cortex [89, 157, 346].

We thus conclude that multiple source modeling of SEPs on scalp-EEG is useful to study dynamical functional anatomy of human somatosensory cortex non-invasively. Furthermore, this technique might become an easy-to-use, non-expensive aid in patients undergoing neurosurgery adjacent to central fissure to delineate essential cortex, and thus could help to avoid disabling neurologic deficits in these patients. These issues will be further discussed in Section 6.

4.3.3. Median Nerve Somatosensory Evoked Fields on Magnetoencephalography

4.3.3.1. Data

As already mentioned in the Introduction (Section 1.2.3.4) somatosensory evoked magnetic fields (SEFs) can be recorded over contralateral primary somatosensory cortex using either a steady state paradigm [48, 254, 268, 342] or transient evoked responses [34, 42, 166, 176, 197, 200, 222, 309, 311, 346, 352, 353, 375]. The aim of the present study was to analyze functional anatomy of human somatosensory cortex, i.e. the spatiotemporal structure of sequential sensory information processing. Therefore, we used transient evoked responses as they represent a more realistic model of sensory volleys arriving at somatosensory cortex.

Concerning transient SEFs, studies focusing on short latency and mid-to-long latency responses can be distinguished. Many previous MEG studies reported on mid-to-long latency SEFs, i.e. on the latency range between 30 and 200 msec [166, 176, 197, 200]. We were mainly interested in early magnetic fields up to a latency of about 40 msec which reflect activity in primary somatosensory cortex, whereas later fields probably get contributions from secondary sensory or association areas. Many previous MEG studies on SEFs used low passband filter settings and sampling frequencies [166, 176, 197, 200]. These recording strategies, however, can obscure early components and small latency differences and furthermore result in artifactual latency shifts [102, 233]. We used a wide passband from 1 to 1000 Hz and a sampling frequency of 4000 Hz. Thus, our results should be compared directly only to studies using similar filter settings and sampling frequencies [346, 352, 353, 375].

Similar to previous studies on short latency SEFs, we could record the first magnetic cortical activity at about 20 msec which will be referred to as SEF_{20} in the following. Other authors have labeled this component as N20 [309, 346, 375] or N19m [352, 353]. In agreement with these reports, the magnetic field pattern in the present study showed a reversal of orientation between the upper and lower ends of central sulcus emerging from the head superior and re-entering the head inferior. Rossini et al. [309] reported on dipolar magnetic field patterns occurring as early as 15 msec. The field extrema were further apart than for the N20 component suggesting activity in deeper structures corresponding to thalamocortical afferents. We could not record consistent magnetic activity in this latency range. This is in agreement with other studies [346, 352, 353, 375] and can be explained by the fact that the second-order gradiometers used in the present study are sensitive preferably to cortical activity whereas they are relatively insensitive to subcortical activity [162].

The next activity which we could record was a rather small component occurring at about 25 msec and was labeled as SEF_{25}. In the raw data, this component was not always clearly discernible. Sutherling et al. [346] could identify this component in some, but not in all patients and referred to it as P25. Tiihonen et al. [352, 353] observed a similar small component in the latency range of 20.3-23.0 msec and called it P22m. Lim et al. [222], using a single channel system in an unshielded environment, found P22m components in all subjects investigated. In this study, the magnetic field strengths of the N19m and P22m components were comparable and the magnetic field patterns were of reversed directions for these components in 3 out of 4 subjects. These authors attributed the P22m to radial neuronal activity in precentral gyrus although radial activity is not supposed to generate magnetic fields in a spherical volume conductor [150]. Furthermore, the results of such a prominent P22m activity are in contrast to most other MEG studies [34, 42, 346, 352, 353, 375] and therefore should be viewed with caution.

In our study, the next and thus second prominent peak occurred at 30 msec and was termed SEF_{30}. The waveforms reported by different groups after the SEF_{20} and SEF_{25} components differ considerably concerning latency and field pattern. This can be attributed to different sampling frequencies, recording bandwidths, interstimulus intervals, and recording equipment. Some authors refer to this activity as P27m component although its latency is around 30 msec [352, 353], other groups have termed it P30 component [34, 42, 346, 375]. In our study, SEF_{30} was associated with a magnetic field pattern which reversed polarity between the upper and lower ends of central sulcus, and was of opposite orientation compared to the SEF_{20}, i.e. emerged from the head lateral inferior and re-entered the head medial superior. Although the magnetic field pattern associated with the SEF_{30} component was similar to that of the SEF_{20}, the two field patterns were not identical. These findings agree with most of the aforementioned studies [346, 352, 353, 375].

Thereafter, we could observe a rather small activity at about 35 msec referred to as SEF_{35} component. Reports in the literature on this component are scare and even more divergent [176, 222, 311, 346]. In previous studies, it has been suggested that magnetic activity in the primary hand projection area continues until about 150 msec after median nerve stimulation [166].

The effects of interstimulus intervals (ISI) on SEF waveforms are controversial. Some authors believe that changing the ISI affects various

SEF components differently. Tiihonen et al. [352] found that increasing the ISI increased the amplitude of the P27m component more than that of the N20m component and therefore suggested contributions of different neuronal networks to these components. We tested different ISI in some subjects and found only minor effects on amplitude and virtually no effect on the spatiotemporal magnetic field pattern. The effect of different ISI thus seemed to be negligible and we therefore used a constant stimulus repetition rate of 3.1 Hz in all subjects.

4.3.3.2. Spatiotemporal Modeling

Our results suggest that MEG is a useful tool to study functional anatomy of human somatosensory cortex non-invasively. Furthermore, spatiotemporal modeling of the entire time domain of SEFs with multiple dipoles fixed in space and varying in activity over time enabled us to study the number, three-dimensional location, and time activity of the neuronal sources generating human somatosensory evoked magnetic fields to median nerve stimulation. We found that two dipolar sources were sufficient to model the first 60 msec of the SEFs accounting for 86.4% of the data variance. Adding a third source did not significantly increase the variance explained by the model. Both sources were located close to the C3/C4 position of the International 10-20 System, which is reasonable since the distance between C3/C4 and central fissure is usually assumed to be less than 15 mm [173, 174, 180, 331]. Thus, both sources were located in or close to primary hand somatosensory cortex. Both sources showed overlapping time activities. One source generated the SEF_{20}-SEF_{30} component, and the other source generated the SEF_{25}-SEF_{35} component. The SEF_{20}-SEF_{30} source had peak latencies at 21 and 31 msec, and the SEF_{25}-SEF_{35} source exhibited peak latencies at 24 and 34 msec. Multiple dipole modeling could separate the spatiotemporal magnetic field patterns of the neuronal networks underlying SEFs which overlapped both in space and time and revealed an underlying simplicity of the human somatosensory evoked magnetic response.

Our approach of multiple dipole modeling extends previous work on somatosensory evoked magnetic fields (SEFs). So far, only single moving dipole models have been used to study SEFs. Furthermore, these studies restricted source localization to selected latencies given by peak amplitude or constancy of field distribution [166, 176, 200, 222, 254, 268, 346, 352, 353, 375]. The spatiotemporal model proved to be superior to these approaches especially for activity occurring after 20 msec, i.e. for interpretation of the SEF_{25} and SEF_{30} components because in this latency range SEFs are generated by multiple simultaneously active brain areas [8, 32, 34, 42, 346, 375].

Clear identification of the SEF_{25} component from visual analysis of the data seems to be inconsistent across subjects as can be seen in the paper of Sutherling et al. [346] and from our raw data. In most previous SEF studies using single equivalent dipole models, the source of the SEF_{25} component could not be determined due to its small size and its superimposition on the SEF_{20}-SEF_{30} slope [346, 352, 353]. The spatiotemporal model thus was superior to these single moving dipole models as it could clearly separate the activities arising from the SEF_{20}-SEF_{30} source and from the SEF_{25}-SEF_{35} source, and thus enabled us to determine the three-dimensional location of the SEF_{25}-SEF_{35} source. A possible explanation for these differences might be that spatiotemporal modeling could have extracted the separate covariance structures of two different sources which overlapped in space and time. This would be expected from previous theoretical work with simulations of multiple dipoles [40].

Isofield maps for the SEF_{30} component show similar, but slightly different magnetic field patterns compared to those of the SEF_{20} component. Therefore, several authors have suggested separate generators for these two components. On the contrary, spatiotemporal modeling yielded one source generating both the SEF_{20} and the SEF_{30} component. Thus, our results suggest that these differences in magnetic field patterns can be accounted for by overlapping activity from the SEF_{25}-SEF_{35} source and not by different sources. These issues will be further discussed in Section 4.3.3.3.

4.3.3.3. The Neuronal Sources Underlying SEFs

Our working hypothesis is that early SEFs are generated by two neuronal sources, one generating the SEF_{20}-SEF_{30} component, and the other generating the SEF_{25}-SEF_{35} component. We believe that the source underlying the SEF_{20}-SEF_{30} component is generated by a tangential dipole in the posterior wall of central sulcus corresponding to Brodmans's area 3b for several reasons. First, the SEF_{20}-SEF_{30} source was located at an average depth of 9 mm below the cortical surface which is consistent with activity at the centroid of central sulcus which has been found to be 18 mm deep [8]. Additionally, the SEF_{20}-SEF_{30} source was located deeper than the SEF_{25}-SEF_{35} source which can be regarded as a depth marker for cortical surface activity. Second, the SEF_{20}-SEF_{30} source was oriented primarily in the anterior-posterior direction resulting in a magnetic field pattern which reversed orientation between the upper and lower recording positions. Synchronized synaptic activity at the cell bodies and the proximal apical dendrites of cortical pyramidal cells in area 3b generates an intracellular current from the cell bodies to the distal dendrites, i.e. a current directed anterior towards the central fissure. As MEG selectively measures intracellular currents [23, 26, 265, 266, 267], this neuronal activity would result in a magnetic field pattern emerging from the head medial superior and re-entering the head lateral inferior as observed for the SEF_{20} component. On the contrary, synchronized synaptic activity at the distal apical dendrites generates an intracellular current from the dendrites towards the cell bodies, i.e. a current directed posterior and away from central sulcus, and results in a magnetic field pattern emerging from the head lateral inferior and re-entering the head medial superior as observed for the SEF_{30} component. These findings of identical or similar neuronal sources of SEF_{20} and SEF_{30} and therefore sequential activation of proximal and distal apical dendrites at cortical pyramidal cells are in agreement with the results on ECoG (Sections 4.2.1.4 and 4.3.1.5) and scalp-EEG (Sections 4.2.2.3 and 4.3.2.4) in this study.

Whereas there is general agreement from most previous MEG studies that the SEF_{20} component is generated by activity in area 3b [34, 42, 162, 166, 176, 200, 222, 309, 346, 352, 375], the neurogenesis of the SEF_{30} component is more controversial. Wood et al. [375] found similar, but not identical magnetic field patterns for SEF_{30} compared to SEF_{20} and concluded that SEF_{30} gets a small contribution from radial sources in somatosensory cortex or motor cortex, or both. Huttunen et al. [176] as well as Tiihonen et al. [352, 353] found that the sources underlying the SEF_{30} were located at an average of 10 mm anterior and medially to those of the SEF_{20} component and that the SEF_{30} component was affected differently by changing interstimulus intervals. Therefore, these authors concluded that SEF_{30} gets a contribution from the anterior wall of central sulcus. On the contrary, Sutherling et al. [346] found no consistent differences between the SEF_{20} and SEF_{30} sources concerning the anterior-posterior or medial-lateral direction. As already mentioned in Section 4.3.3.2, we believe that these different source localizations for the SEF_{20} and SEF_{30} components are the result of modeling activity generated by multiple brain regions with a single equivalent dipole which can lead to artifactual localization shifts [40, 259]. Thus, our interpretation is in line with that of Wood et al. [375]. Furthermore, we conclude from these findings that spatiotemporal modeling can enhance non-invasive localization accuracy on MEG.

Rossini et al. [311] found two additional extrema on both sides of the head during the SEF_{30} component. Thus, these authors suggested additional activation of frontal parasagittal midline structures corresponding to non-primary motor cortex. Our measurement matrix extended up to 7.5 cm anterior to the C3/C4 position, but we did not record over the contralateral hemisphere. We could not observe such activity and a different recording strategy would be necessary to infer this question systematically.

Concerning the SEF_{25}-SEF_{35} source, our hypothesis is that this activity is generated by radial activity in the anterior crown of postcentral gyrus corresponding to area 1 for several reasons. The SEF_{25}-SEF_{35} source was located more superficially compared to the SEF_{20}-SEF_{30} source which would be expected from activity near the cortical surface as compared to activity in the depth of the sulcus. Ideally, a radial source does not generate a measurable magnetic field outside a spherical volume conductor [150]. However, combined neuroelectric and neuromagnetic measurements have shown that the dipole generating the electric P25 component on scalp-EEG is tilted about 20° from radial [346]. Thus, the electric P25 seems to have also small tangential components which generate magnetic fields measurable with low-noise equipment [346, 352, 353]. As already mentioned in Section 4.3.3.2, these small components could not be modeled above noise in previous MEG studies with single equivalent dipole models. Spatiotemporal modeling, on the contrary, allowed identification and modeling of this source. The consistently smaller variance contribution of the SEF_{25}-SEF_{35} source in comparison to the SEF_{20}-SEF_{30} source can be explained by the fact that only the tangential components of this primarily radial source were detected in the MEG and therefore also point towards activity in surface cortex of area 1. Nevertheless, it should be kept in mind that magnetic recordings are blind to radial dipoles and therefore no conclusions can be made about the radial components from our results.

In conclusion, we think that from neuromagnetic measurements alone the exact anatomical location of the two sources cannot be determined with certainty due to the lack of a constant relative location between the two sources and between the individual sources and the C3/C4 position of the International 10-20 System. Furthermore, the relation of anatomical landmarks on the skull to cytoarchitectonic brain structures is variable [180, 331] and functional organization underlies interindividual variations. Nevertheless, MEG together with ECoG and scalp-EEG yielded converging evidence towards a unified hypothesis concerning the neurogenesis of the human somatosensory evoked response. These issues will be discussed in further detail in Section 4.3.6.

4.3.3.4. Functional Organization of Human Somatosensory Cortex

While we found considerable latency differences between the peaks of the SEF_{20}-SEF_{30} source and the peaks of the SEF_{25}-SEF_{35} source in some subjects (cf. Subject #2), latency differences were very small in other subjects (cf. Subject #1). There are alternative explanations of these smaller latency differences. The first possibility is that such a short latency between two similar waveforms could arise because the spatiotemporal model represented one brain region artifactually as two separate sources. This appears unlikely for several reasons. The sampling frequency was adequate to avoid temporal aliasing. Principal component analysis (PCA) identified two independently covarying parts underlying the data. The additional variance explained by the SEF_{25}-SEF_{35} source was larger than run-to-run variability. Furthermore, the SEF_{20}-SEF_{30} source was located consistently deeper compared to the SEF_{25}-SEF_{35} source in all subjects, including those who had longer intersource latencies. The second possibility is that the spatiotemporal model could have separated two sources with overlapping fields which were not distinguishable using either visual inspection of waveforms and map features, or using single moving dipole models. Since the first alternative is unlikely, the second alternative remains as our working hypothesis.

Thus, the SEF_{20}-SEF_{30} source and the SEF_{25}-SEF_{35} source could be activated either in parallel by separate thalamocortical projection systems or sequentially by synaptic propagation via cortico-cortical connections. The difference of 1.4 msec for peak activity (cf. Subject #1) appears very fast for synaptic propagation between different brain regions. In this Subject, the distance between the two sources, or the centers of cortical regions was

9 mm. Since the actual distance over U fibers in the cortex would be longer due to the anatomical curvature of fibers in gyri, the dipole estimate of 9 mm is likely the lower limit of the distance traveled. Measured cortico-cortical conduction velocities vary between 2.5 and 28 m/sec but are usually 2.5 to 8 m/sec [350, 364]. Using the lower limit for distance and the upper limit for usual conduction velocity, this gives a transit time of 1.13 msec. Although the smallest measured synaptic delay in the mammalian nervous system is 0.2 msec [47], synaptic delay is usually about 0.5 to 0.9 msec [312]. Most quantitative studies on cortico-cortical and synaptic propagation have been performed in animals; however, propagation times seem to be very similar in human neocortex as evidenced from studies in epileptic patients [344]. Therefore, the lower limit for synaptic propagation time between the centers of two human neocortical regions separated by 9 mm would be 1.63 (1.13 + 0.5) msec to 2.03 (1.13 + 0.9) msec. This is longer than the measured latency difference between the time activity of the SEF_{20}-SEF_{30} source and that of the SEF_{25}-SEF_{35} source. If one assumes that the relative time activity accurately reflects the absolute latencies between individual neurons in the two cortical regions, then the measured latency difference is smaller than the expected latency difference if there were serial activation of the two sources by synaptic propagation of activity from the SEF_{20}-SEF_{30} source to the SEF_{25}-SEF_{35} source. Thus, our findings favor the alternative possibility that the two sources are activated in parallel by volleys from thalamocortical fibers. This hypothesis is consistent with the results from animal studies demonstrating separate projection systems from thalamus to areas 3b and 1 [188]. It is certain that the large majority of the activity of the two sources does occur in parallel. This study shows the largest overlap of the activity of the two sources in the literature to our knowledge. Therefore, we conclude that spatiotemporal modeling may be useful to form testable hypotheses about the functional anatomy of human hand somatosensory cortex.

4.3.4. Comparison of ECoG, Scalp-EEG, and MEG

4.3.4.1. Comparison of ECoG and Scalp-EEG

Concerning the comparison of the raw data, we could observe a striking similarity between SEPs on scalp-EEG and ECoG. Characteristic waveforms were almost identical (Fig 4.2.4.1). Cortical surface recordings were of larger amplitude and had a restricted spatial pattern compared to scalp-EEG. This finding can be attributed to the resistive properties of skull and scalp. As already noted by DeLucchi et al. [96], the skull acts as an electroencephalographic averager and only synchronized cortical activity shows up at the scalp, whereas local activity tends to cancel out. Thus, from our findings we conclude that SEPs recorded on the cortical surface are mainly composed of highly synchronized neuronal activity, and that most of this information also can be obtained at the scalp.

Our findings are in agreement with those of previous investigators. Lüders et al. [226] found similar waveforms for SEPs on ECoG and scalp-

EEG (cf. Fig. 1 in Lüders et al. [226]). As main differences between ECoG and scalp-EEG, these authors noted the large amplitude and the extremely localized spatial pattern of cortical recordings. Concerning amplitude, Dinner et al. [113] studied postrolandic cortical SEPs in 12 epileptic patients and observed an N1-P2 amplitude of 35.3 µV (range 8.6-85.7 µV), which was 8.7 times (range 3.0-15.7) larger than the corresponding scalp-potentials. We could observe a mean attenuation factor of 11.2. Similar observations were made by Allison et al. [4]. These authors noted similarities in the wave morphologies of the raw data and potential fields obtained from the scalp and from the cortical surface. Concerning the spatial distribution, there were large differences between cortical and scalp recordings in the location of amplitude maxima of these potentials. Cortical surface recordings were large (50-100 µV) and had steep spatial gradients confined to the region of the central sulcus. Scalp-EEG recordings, on the contrary, showed amplitude peaks near P4 and F4, well posterior and anterior to sensorimotor cortex. On the contrary, Sutherling et al. [346] observed significant differences in the raw data between scalp-EEG and ECoG. Whereas EEG showed a widespread pattern, ECoG was more complex showing both a widespread pattern similar to scalp-EEG and furthermore a restricted pattern with additional peaks that did not show up at the scalp. These authors attributed these differences to distance effects.

Our findings extend previous studies as we also compared the results of spatiotemporal modeling obtained from ECoG and scalp-EEG. We found several striking similarities between the results obtained with these two recording techniques. First, both on ECoG and scalp-EEG, the modeling procedure yielded two neuronal sources, namely one source generating the N20-P30 component, and the other generating the P25-N35 component. Second, the spatial potential distributions of the N20-P30 source showed tangential patterns with a phase reversal across central fissure both on ECoG and scalp-EEG. On the contrary, the spatial patterns for the P25-N35 source were different on ECoG and scalp-EEG. Whereas on scalp-EEG this source consistently showed a radial pattern in all subjects, on ECoG the orientation of this source was more variable, i.e. in some patients it was more tangential (cf. Patient #1), and in others it was more radial (cf. Patient #2). Third, on ECoG the depth estimates for the two sources were similar. On scalp-EEG, on the contrary, the N20-P30 source was localized deeper compared to the P25-N35 source. These findings for the P25-N35 source on ECoG, i.e. significant tangential components and similar depth estimates compared to the N20-P30 source, could be explained by inaccuracies of the dipole concept when modeling cortical surface activity in area 1. As already mentioned in Section 2.2, for this situation one of the model assumptions of dipole modeling is violated, i.e. that measurements are made at a distance which is large compared to the dimensions of the source. These discrepancies warrant further investigation. Fourth, the time activities of both sources showed very good agreement for ECoG and scalp-EEG which further enhances biological reliability of our results. Thus, we conclude that both ECoG and scalp-EEG can be used to study functional anatomy of human hand somatosensory cortex and that the basic conclusions from both techniques are similar.

However, evidence exists from the studies of Sutherling et al. [346] that the relationship between ECoG and scalp-EEG is not always as simple as in the patient presented. Certainly, we did not study enough patients to conclude on these ambiguities. Therefore, further studies are needed to investigate possible factors influencing the ECoG – scalp-EEG relationship.

4.3.4.2. Comparison of Scalp-EEG and MEG

Scalp-EEG was more complicated than MEG. Whereas scalp-EEG showed activity already preceding the N20 component thus representing subcortical activity, virtually no activity could be seen on MEG during this time interval (Figs. 4.2.4.4 and 4.2.4.6). These differences can be explained by the fact that early activity on scalp-EEG arises from subcortical structures. For our MEG studies, we used a second-order gradiometer which is sensitive to cortical activity, but relatively insensitive to activity arising from deep structures [162].

An even more striking difference between scalp-EEG and MEG could be observed for activity between 20 and 30 msec where scalp-EEG showed additional peaks, i.e. the P25 component. These findings are similar to the studies of Wood et al. [375] and of Sutherling et al. [346] who systematically compared scalp-EEG and MEG.

Our study extends previous studies, as we decomposed scalp-EEG in its contributions from tangential sources (N20-P30 source) and from radial sources (P25-N35 source). Comparison of the tangential EEG components with the MEG raw-data yielded striking similarities between these waveforms (Fig. 4.2.4.4). Thus, a physiological hypothesis concerning the neuronal generators of the human somatosensory evoked responses, i.e. tangential activity in area 3b and radial activity in area 1, and a physical hypothesis concerning the neurogenesis of MEG, i.e. that MEG detects primarily tangential dipoles – could be proven to mutually support each other.

Furthermore, we compared the results of spatiotemporal modeling obtained on scalp-EEG and MEG. The orthogonal spatial patterns of the electric N20-P30 source and the magnetic SEF_{20}-SEF_{30} source are consistent with a single neuronal generator in area 3b. Furthermore, the distance of magnetic field maxima was about 40.8 % smaller than that of the electric potential maxima. These results of orthogonality and a more restricted pattern on MEG compared to scalp-EEG are consistent both with previous SEP studies on MEG [346, 375] and with theoretical considerations [69, 81]. The depth estimates for both the electric N20-P30 source and the magnetic SEF_{20}-SEF_{30} source were similar and comparable to reports in the literature [176, 200, 268, 353]. The dissimilarities concerning spatial and temporal pattern of the electric P25-N35 and the magnetic SEF_{25}-SEF_{35} source can be explained by the fact that P25-N35 represents the entire activity in area 1, whereas SEF_{25}-SEF_{35} only represents its tangential component. Nevertheless, depth estimates for this source were comparable.

4.3.4.3. Localization Accuracy of Scalp-EEG and MEG

In our study, we could compare localization estimates only for scalp-EEG and MEG. We found good agreement of these localization estimates concerning surface projection. This is in agreement with the landmark study of Sutherling et al. [346] who compared localization estimates on scalp-EEG, MEG, and ECoG. These authors found that the combination of all non-invasive estimates from scalp-EEG and MEG had localization accuracies approaching that of ECoG. However, these authors obtained depth estimates of 4 cm below the scalp on EEG, which was deeper compared to MEG. Some previous studies predicted that depth localizations should be more accurate in MEG than in scalp-EEG, as in MEG no correction factors for skull and scalp resistance are needed [369, 370]. These findings and theoretical suggestions are in contrast to our results because we achieved good agreement of depth estimates on scalp-EEG and MEG. In this context, also localization accuracies obtained for both interictal as well as ictal epileptiform discharges should be mentioned. In a recent study, MEG was accurate within 12 mm in localizing a stereotyped well circumscribed epileptic spike focus with maximal amplitude consistently at one electrode on ECoG [345]. Furthermore, good agreement could be obtained between MEG and ECoG for localization of ictal epileptiform discharges [347]. We thus conclude that the combination of scalp-EEG and MEG should help to improve non-invasive localization of physiological as well as pathological brain activity. Whether the scalp-EEG and MEG combination will also be useful in clinical routine for exact localization of epileptiform activity arising from mesial temporal lobe in patients with complex partial seizure remains to be awaited. Certainly, further progress in MEG technology like planar SQUIDs, multichannel devices, and gradiometers sensitive to activity arising from deep brain structures, i.e. first order gradiometers would be helpful in this direction.

4.3.4.4. Additional Information Revealed by Scalp-EEG and MEG

N20 and P30 showed a phase reversal across central sulcus. This situation could arise either from one horizontal dipole in the posterior wall of central sulcus or by two radial dipoles of opposite polarity in pre- and postcentral gyrus, respectively. Allison et al. [4] investigated the single and the dual source hypothesis in a model study. Both models were adjusted to produce the observed highly focal cortical surface potentials whose positive and negative peaks were located about 8 mm anterior and posterior to central sulcus. Under these conditions, the single source model predicted scalp amplitude peaks in close proximity to P4 and F4 and thus in good agreement with the actual data measured on scalp-EEG. For the dual source model, on the contrary, scalp amplitude peaks were located much closer to each other. In order to produce the observed scalp topography with the dual source model, the sources had to be moved anterior and posterior to the F4 and P4 position and then were incompatible with the cortical surface potentials. Thus, the combination of ECoG and scalp-EEG could resolve an ambiguity concerning the neurogenesis of the N20 component.

Wood et al. [375] and Sutherling et al. [346] have proven that MEG can further clarify this situation. As already mentioned, MEG measures selectively tangential activity in the sulci [150]. At 20 and 30 msec, tangential magnetic field patterns orthogonal to the electric field patterns could be observed. These findings can be only explained by a single horizontal source as two radial sources could not possibly account for the observed electric and magnetic field patterns simultaneously. The theoretical foundations for this reasoning have been presented in detail in Section 2.3.4, implications concerning the neurogenesis of the human evoked response are further discussed in Section 4.3.6.4. Thus, MEG helped to clarify and complemented the information of cortical surface and scalp-EEG recordings. MEG may also be helpful in still more complicated situations such as the epileptic focus to distinguish between a focal or widespread seizure onsets [347].

4.3.5. Considerations Concerning the Model

4.3.5.1. The Dipole Concept

Organization of human primary motor and somatosensory cortex as revealed by anatomical and physiological studies is complicated by distinct and independent afferent projections from thalamus [182, 185, 187, 188], distinct complete representations of hand in both sensory [192, 193, 194, 244, 245] and motor cortex [339, 340, 341], and several intracortical connections between sensory and motor areas [183, 184, 186]. Therefore, it may be an oversimplification to model these complicated interactions by simple current dipoles. However, as already mentioned in Section 2.2, the dipole concept is justified for several reasons. First, the electric potentials recorded on scalp-EEG and ECoG as well as the magnetic fields recorded on MEG are generated by synchronized synaptic activity of the cortical pyramidal cell bodies and their apical dendrites which are organized in columns perpendicular to the surface of the cortex [223]. Synchronized synaptic activity of these cells thus generates current dipoles. Second, incoming thalamic afferents result in activation of extended areas of somatosensory cortex rather than in simple point sources. However, even these more complex sources, i.e. dipole layers or sheets representing patches of activated cortex can be modeled well by simple dipoles without introducing significant errors in localization as has been shown in simulation studies [89]. Thus, it is the center of neuronal activity rather than its spatial extent that is estimated by dipole modeling. Therefore, dipoles should be viewed a useful mathematical abstractions rather than as real neuronal elements [302]. Finally, the electric and magnetic field generated by each current source with equal number of positive and negative charges falls off similar to that induced by a current dipole when measurements are made at a distance which is large compared to the dimensions of the source. This can be explained by the fact that the higher order terms fall off with distance more rapidly [162, 258]. These conditions are fulfilled for scalp-EEG, MEG, and with some restrictions also for ECoG. In certain circumstances, however, this assumption may cause minor problems, especially when cortical surface activity is modeled on ECoG. This can be seen from the depth estimates obtained for the P25-N35 source on ECoG.

4.3.5.2. Spatiotemporal Modeling – Modeling Assumptions and Neurophysiological Considerations

The human somatosensory evoked response is generated by multiple simultaneously active brain areas [7, 8, 32, 42, 193, 375]. The application of single dipoles to model this activity generated by multiple brain regions may yield misleading results [40, 259]. Spatiotemporal modeling, on the contrary, was useful to separate the spatiotemporal field patterns of the sources underlying the evoked response, and to investigate their three-dimensional intracerebral locations as well as their time activities. Thus, an underlying simplicity of the sources producing complex spatiotemporal field patterns could be revealed.

A limitation of our model may be the assumption of spatially fixed sources with time-varying activities. Although anatomically fixed sources appear reasonable for modeling the human somatosensory evoked response, this does not allow modeling of rapidly shifting intracortical propagation. It is unlikely that this is a major practical limitation, since only highly synchronized synaptic events produce organized activity measurable on scalp-EEG, MEG, and with some restrictions also on ECoG. As already mentioned, DeLucchi et al. [96] have shown that only highly synchronized brain activity can be measured on scalp-EEG whereas local activity cancels out due to the resistive properties of the skull. Thus, the skull can be regarded as an electroencephalographic averager. Similarly, for MEG it has been shown that due to the thickness of skull, scalp, and the insulating dewar of the magnetometer, spatially averaged activity is measured [82, 304] and only highly synchronized synaptic activity in a large neuronal pool is detected [23, 26, 265, 266]. Finally, our results on ECoG have demonstrated that the basic qualitative features of the evoked responses are similar on ECoG compared to scalp-EEG – namely similarity of the raw data and similar results on spatiotemporal modeling. This is agreement with previous reports on the ECoG and scalp-EEG comparison [4, 7, 51, 113, 226]. These findings, therefore, validate the modeling assumptions for ECoG as well. Thus, it is the center of mass of these neuronal populations that is estimated by source localization methods rather than rapidly shifting propagation within local neuronal networks. Finally, it seems reasonable from a neuroanatomic viewpoint to assume spatially fixed sources for some applications rather than moving dipoles since neuronal populations do not move in the brain. For special questions concerning propagation patterns in somatosensory cortex, however, different models could be more appropriate.

4.3.5.3. Spatiotemporal Modeling – Mathematical and Computational Considerations

Spatiotemporal modeling is a delicate task from the computational standpoint. The modeling procedure involves a non-linear minimization algorithm to estimate the dipole parameters. Non-linear minimization necessitates numerical techniques as there exist no explicit analytic solutions. These minimization procedures have to be provided with initial estimates of the dipole parameters, the so-called start values. In the present study, we used the simplex algorithm introduced by Nelder and Mead [255, 295]. This algorithm has two main advantages. First, it is very stable from a mathematical viewpoint. This means that changing the start values generally does not greatly affect the results. Second, it is not necessary to provide gradients of the function to be minimized [295]. This is very convenient for the electrophysiological inverse problem as it is very cumbersome to provide explicit formulations of the gradients.

Nevertheless, several precautions have to be taken into account when using the simplex algorithm for spatiotemporal modeling. In theoretical studies, it has been shown that – despite the general robustness of the simplex algorithm – spatiotemporal modeling can prove as a highly ill-conditioned problem. Thus, different start values yield totally different solutions, i.e. dipole parameters, due to the existence of local minima [40, 215]. Therefore, special care should be taken to a proper selection of start values. Ideally, the investigator should provide some a-priori knowledge concerning the underlying anatomy and physiology. This condition was fulfilled for the study of the somatosensory system where extensive a-priori knowledge exists both from human and animal studies. Furthermore, we developed a method using principal component analysis which allowed selection of start values based on objective criteria of the covariance structure of the evoked response [40], see also Sections 3.5 and 3.7. Furthermore, it should be mentioned that the different solutions obtained by different start values often reproduce the actual data equally well, i.e. account for virtually the same amount of data variance. Thus, a high percentage of variance accounted for is not equivalent with the correctness of the solution [40, 215]. Another possibility to deal with these problems is a new minimization procedure called simulated annealing. This procedure was initially proposed by Kirkpatrick et al. [207] and is based on an analogy with thermodynamics. The advantage of this method is that it achieves the global minimum beyond some local minima, i.e. it provides the correct solution in terms of minimization. This procedure was tested in simulation studies on the biomagnetic inverse problem where it could be shown that with one or two dipoles there was no difference between simulated annealing and conventional minimization procedures [159]. On the other hand, with three or more dipoles only simulated annealing provided the correct solution.

In conclusion, spatiotemporal modeling may be considered a useful method to form testable hypotheses about brain physiology. However, the final decision on plausibility and correctness of the results is based on a sound physiological knowledge of the investigator.

4.3.6. Neurogenesis of the Human Somatosensory Evoked Response

4.3.6.1. The Primary Evoked Response

In animal experiments, the first neural activity in primary receiving areas in response to an afferent volley from the corresponding thalamic relay nucleus is a surface positive-negative waveform, usually referred to as primary evoked response. The primary evoked response has been extensively studied in in various species, i.e. monkeys, cats and other mammals, and in various modalities, i.e. in the visual, the auditory, and the somatosensory modality [6, 246, 327, 356]. Concerning its neuronal substrates, the positivity is thought to be generated by the initial synaptic depolarization at the cell bodies and the proximal apical dendrites of cortical pyramidal cells, whereas the negativity is thought to be generated by the later depolarization of the distal apical dendrites [78, 212, 327, 365, 373].

Allison et al. [8] suggested that in human primary somatosensory cortex analogous surface positive-negative potentials are generated in area 3b (P20-N30 components) and a few milliseconds later in area 1 (P25-N35 components) in response to an incoming somatosensory volley and thus generate the human primary evoked response. In monkeys, similar observations could be made as the primary response in area 3b occurred a few milliseconds earlier than in area 1 [10, 14, 391]. These differences in response onset times in areas 3b and 1 can be explained by separate projection systems from the ventral posterior lateral nucleus of the thalamus [188]. Whereas input to area 3b consists of a dense projection of fast conducting large-diameter fibers, area 1 receives a sparse input from slower conducting smaller fibers [188, 294, 358].

Our results on spatiotemporal modeling agree with these hypotheses. With three different recording techniques – electrocorticography, scalp-EEG, and magnetoencephalography – we could identify two dipolar sources which overlapped in space and time and explained the large majority of the human somatosensory evoked response. One source generated the N20-P30 component and was consistent with activity in area 3b, and the other generated the P25-N35 component and was compatible with activity in area 1. In the following, activity attributed to specific cytoarchitectonic structures will be discussed in detail.

4.3.6.2. Neurogenesis of the N20-P30 Component – Activity Attributed to Area 3b

Our findings suggest that the N20-P30 components on ECoG and scalp-EEG, and the SEF_{20}-SEF_{30} components on MEG are generated by a tangential dipolar source in the posterior wall of central sulcus corresponding to area 3b. This hypothesis is in agreement with the findings of other investigators [4, 7, 8, 346, 375]. An alternative hypothesis asserts that P20-N30 and N20-P30 are independently generated in the crowns of motor and sensory cortex [275, 276]. The following points strongly favor the first hypothesis, i.e. that the N20 and P30 are generated in area 3b.

1. Evidence from Spatiotemporal Modeling. Spatiotemporal modeling of evoked responses obtained with three different techniques yielded a single tangential dipolar source in postcentral gyrus underlying the N20-P30 component. This technique allows to analyze the entire time domain of the evoked response and to infer – besides the three-dimensional localizations of the sources – also their time activities. The hypothesis that the N20-P30 is generated by two distinct neuronal populations in motor and somatosensory cortex would require that these two sources have identical time activities over the entire evoked response. Animal studies have documented direct input from thalamus both to area 3b and to area 4. However, the projections systems from thalamus to motor and somatosensory cortex differ in fiber diameters resulting in different conduction velocities [16, 17, 18, 182, 183, 185, 187, 188]. Thus, an exactly time locked synaptic activation in motor and somatosensory cortex seems extremely unlikely.

2. Evidence from Source Localizations. On ECoG, the N20-P30 source was located in postcentral gyrus at an average distance of 6 mm from central fissure. Its surface projection was close to electrodes which elicited sensory experiences in the hand during cortical stimulations. These findings were reproducible across 6 patients and therefore suggest a true postcentral N20-P30 source. Furthermore, scalp-EEG and MEG yielded supportive evidence. Scalp-EEG and MEG source localizations were at an average distance of 9 mm from the C3/C4 position of the International 10-20 System and therefore close to central fissure as C3/C4 is within 15 mm of central fissure [173, 174, 180, 331]. Scalp-EEG and MEG, however, do not allow a clear attribution of the N20-P30 source to either pre- or postcentral gyrus.

3. Evidence from Source Orientations. Spatiotemporal modeling on three different recording techniques yielded N20-P30 sources which were oriented tangentially and approximately perpendicular to central sulcus. These findings support the hypothesis of activity in the wall of postcentral gyrus corresponding to area 3b where neurons are arranged in columns perpendicular to the sulcus and parallel to the cortical surface [293, 361].

4. Evidence from Depth Estimates. The depth estimates of the N20-P30 source as provided from spatiotemporal modeling of evoked responses on ECoG, scalp-EEG, and MEG showed good agreement, i.e. 8 mm on ECoG, 10 mm on scalp-EEG, and 9 mm on MEG. The human central sulcus is about 18 mm deep [8]. Thus, the depth estimates for the sources are consistent with activity at the center on either wall of central sulcus. These depth estimates, however, are not consistent with activity in the surface cortex either pre- or postcentrally. The human primary somatosensory cortex averages a thickness of 2.5-3 mm [281] and during the primary potentials current sources and sinks are seen in the upper and middle cortical layers approximately 1 mm below the surface [124, 246, 356, 365].

5. Evidence from Combined Use of Neuroelectric and Neuromagnetic Recordings. SEPs on ECoG and scalp-EEG showed prominent N20 and P30 components, which could be explained either by one horizontal dipole in the posterior wall of central sulcus or by two radial sources in motor and somatosensory cortex. However, also SEFs showed prominent peaks at 20 and 30 msec termed SEF_{20} and SEF_{30}. As MEG is primarily sensitive to tangential sources and radial sources are not detected in a spherical volume conductor [150], these results favor the hypothesis of one tangential source. The alternative hypothesis would be that the observed electric and magnetic field patterns were produced by two radial sources in the crowns of pre- and postcentral gyrus which were tilted tangentially to produce non-zero tangential components. However, these sources would produce potential maxima separated by distances 50 to 60 % smaller than those obtained in the scalp-EEG and ECoG recordings [375]. On the contrary, two radial sources located far enough anterior and posterior to account for the potentials obtained on ECoG and scalp-EEG would produce multiple dipole magnetic field patterns which are clearly different from those obtained in the MEG measurements [375]. Thus, combined neuroelectric and neuromagnetic measurements could resolve an ambiguity which could not be decided by either method alone. An analogous combined EEG-MEG approach on this question was used by Sutherling et al. [346] and yielded the same conclusions.

6. Evidence from Scalp, Cortical Surface, Transcortical, and Depth Probe Recordings. As already mentioned, Kelly et al. [204] using transcortical recordings from the human postcentral gyrus found that N20 showed no phase reversal between the cortical surface cortex and the underlying white matter. In agreement with these results, Allison et al. [8] found a phase reversal neither for the precentral P20-N30 components nor for the postcentral N20-P30 components when comparing recordings from the scalp and the cortical surface on the one hand, and from the underlying white matter using depth electrodes on the other hand. These findings thus virtually rule out generation of these components in pre- or postcentral surface cortex.

Further evidence for a generation of these components in area 3b is provided by the following findings of Allison et al. [8]. First, on depth recordings from frontal lobe the largest amplitude of the P20-N30 component was found at a depth of 1-2 cm which would be expected from activity in the central sulcus. Second, the P20-N30 components could be recorded after removal of hand area of motor cortex in a patient with Rasmussen's Syndrome directly from the posterior wall of postcentral gyrus with a maximum amplitude at a depth of approximately 1 cm below the cortical surface which can only be explained by generation of this activity in area 3b.

7. Support from Animal Studies. Finally, these human findings are supported by SEP and multiunit recordings in monkeys [13, 14]. Whereas anterior to central sulcus P10-N20 waveforms similar to the human P20-N30 components could be recorded, posterior to central sulcus N10-P20 waveforms corresponding to the human N20-P30 components could be measured. The shorter latencies can be explained by the shorter conduction pathways. On transcortical recordings, P10-N20 components showed phase reversals across area 3b but not across areas 1 and 2. During P10 there was strong unit discharge in area 3b, but not in area 4. Similar to the results in humans, lesions of motor cortex have little effect on these components, whereas lesions of somatosensory cortex abolish them [10]. Thus, these monkey recordings further support neurogenesis of N20 and P30 in area 3b.

4.3.6.3. Neurogenesis of the P25-N35 Component – Activity Attributed to Area 1

Our findings suggest that the P25-N35 components on ECoG and scalp-EEG and the SEF_{25}-SEF_{35} components on MEG are generated by a radial dipolar source in the anterior crown of postcentral gyrus corresponding to area 1. This hypothesis is in agreement with the findings of other investigators [4, 7, 8, 346]. We believe that this hypothesis is valid for the following reasons.

1. Evidence from Spatiotemporal Modeling. Spatiotemporal modeling on ECoG, scalp-EEG, and MEG evidenced two neuronal sources generating the human somatosensory evoked response. Specifically, a single predominantly radial source generated both the P25 and N35 components. From the arguments presented in the following there is compelling evidence of a significant contribution of area 1 to the evoked response and that significant parts of the P25 and N35 components are generated in this cortical area. The question remains whether there exists a significant contribution from other cortical areas to these components, specifically from area 4. However, both principal component analysis and spatiotemporal modeling indicated two significant covarying brain sources underlying the evoked response, and these have already been attributed to areas 3b and 1. Thus, the only theoretical possibility of an additional significant precentral source is that this source exhibits a time activity exactly identical to that of the N20-P30 source or to that of the P25-N35 source, which is extremely unlikely in view of the varying conduction velocities of thalamic projections to these cortical areas [16, 17, 18, 182, 183, 185, 187, 188].

2. Evidence from Source Localizations. Spatiotemporal modeling of the somatosensory evoked response on ECoG yielded a postcentral location for the P25-N35 source. The surface projection of this source was next to electrodes from which sensory experiences in the hand could be elicited. As these findings were reproducible across 6 patients we think that the location in postcentral gyrus is real and not an artifact. Moreover, scalp-EEG and MEG locations for this source were close to the C3/C4 position and thus within 15 mm of central fissure [173, 174, 180, 331]. However, neither scalp-EEG nor MEG allow attribution of this activity to either pre- or postcentral gyrus.

On ECoG, the P25-N35 source was located at an average of 9 mm medial to the N20-P30 source. This agrees with findings in human ECoG studies, as Wood et al. [378] found that the P25 extrema were 9.9 mm medial to the N20 extrema. This mediolateral displacement of the hand representation in area 1 compared to area 3b is not evident in monkey somatosensory cortex

[194]. However, this finding is supported by other animal studies. Thus, Juliano et al. [191] investigated 2-deoxyglucose labeling in monkey somatosensory cortex during finger stimulation and found two separate, though partially overlapping patches of activity, one in area 3b, and the other approximately 3-4 mm medial in area 1. This distance of 3-4 mm would correspond to a distance of approximately 10 mm between these two areas in human somatosensory cortex [8] and therefore agrees with the distance between the N20-P30 source and the P25-N35 source. Thus, these findings further support attribution of this source to area 1.

3. Evidence from Source Orientations. On ECoG, the orientation of the P25-N35 source was variable across patients. In some patients, it showed a more radial pattern, whereas in other patients it was more tangential. On scalp-EEG, the P25-N35 source had a predominantly radial orientation in all subjects although a slight tilt in the tangential direction could be observed. The orientation obtained from magnetic measurements cannot be used in this context as only the tangential, but not the radial components of this source were measured on MEG. Ideally, radial sources would be expected for activity in the surface cortex of area 1. However, the P25-N35 source probably also gets tangential contributions from activity in the upper portion of the posterior wall of central sulcus which is part of area 1 – besides its predominant location in the anterior half of postcentral surface cortex. In conclusion, the findings of a predominately radial source with a small tangential component are consistent with activity in area 1. The finding of predominantly tangential sources on ECoG could be explained by inaccuracies of the dipole model. One of the basic assumptions of dipole modeling is that measurements are made at a distance which is large compared to the dimensions of the source [258]. For activity in the surface cortex immediately underlying the recording electrode, however, this assumption is violated.

4. Evidence from Depth Estimates. On ECoG, the depth estimates for the P25-N35 source were comparable to those of the N20-P30 source. On scalp-EEG and MEG, the P25-N35 source was located more superficially compared to the P20-N30 source. A more superficial location would be expected for activity in the surface cortex compared to activity in the sulcus and thus the results on scalp-EEG and MEG are reasonable. Concerning absolute depth, the estimates averaged 8 mm on ECoG, 4 mm on scalp-EEG, and 3 mm on MEG. Given an average thickness of 2.5-3 mm for the cerebral cortex [281], these depth estimates for ECoG are to deep even if one takes into account contribution of activity from the upper portion of the posterior wall of central sulcus as a part of area 1. For the ECoG measurements, these discrepancies can be explained by violations of dipole modeling assumptions for measurements from the cortical surface as already discussed in the previous section. For scalp-EEG and MEG, more sophisticated models with realistic geometries (scalp-EEG and MEG) and exact estimates of the thickness and conductivities of skull and scalp (scalp-EEG) would probably be helpful to achieve more accurate depth estimates.

5. Evidence from Combined Use of Neuroelectric and Neuromagnetic Recordings.
Both on ECoG and scalp-EEG, either the N20-P30 or the P25-N35 source
dominated the evoked response. The contribution of a specific source was
variable across subjects. On MEG, on the contrary, the SEF_{25}-SEF_{35} source
yielded only a minor contribution, whereas the SEF_{20}-SEF_{30} dominated the
somatosensory evoked magnetic field. This can be explained by the fact that
MEG picked up only the minor tangential components of predominately
radial activity in area 1. In conclusion, there is converging evidence from the
combination of electric and magnetic measurements that this source is
generated by radial activity in the surface cortex.

6. Evidence from Transcortical Recordings. Transcortical recordings in humans
from postcentral sites near the central sulcus showed surface positive
waveforms in the 23 to 27 msec latency range followed by surface negative
components in the 35 to 40 msec latency range which reversed polarity in
the white matter [145, 204, 335]. These components correspond to the P25-
N35 components in the present study. Furthermore, Kelly et al. [204]
destructed the area from which the response was obtained by ablation of a
core of cortex of 4 mm in diameter extending down to the white matter and
found that the responses were completely abolished. These results therefore
suggest local activity in the anterior crown of postcentral gyrus generating
the P25-N35 components.

7. Support from Animal Studies. On cortical surface recordings in monkeys,
P12-N25 components – corresponding to the human P25-N35 components
– could be recorded from area 1. Transcortical recordings showed a phase
reversal for these components from surface cortex to white matter
indicative of a local generation of these components in area 1.

5. Somatotopy of Human Somatosensory Cortex

5.1. Methods

5.1.1. Somatotopy as Studied with Cortical Stimulations and Somatosensory Evoked Potentials on Electrocorticography

5.1.1.1. Patients

We studied 4 epilepsy patients during presurgical evaluation. All patients had partial seizures defined by EEG recordings and behavior during seizures. The seizures were medically intractable, and the patients were evaluated with chronically indwelling subdural grids for definitive localization of the seizure focus. These patients were a subgroup of the patients in whom functional anatomy of somatosensory cortex was studied. For further details on the patients' clinical seizure characteristics and lesion types see Section 4.1.1.1. All patients had given written informed consent to undergo invasive presurgical evaluation and to participate in the SEP studies.

5.1.1.2. Cortical Stimulations

Cortical stimulations were performed in all patients from chronically indwelling subdural grid electrodes to localize essential cortex. The general principles of this technique are presented in Section 2.4. The special cortical stimulation protocol performed in these patients is described in Section 4.1.1.2.

Furthermore, in these particular patients we were especially interested in somatotopic features of somatosensory cortex. Thus, we paid special attention to the patients' reports when cortical stimulations induced sensory experiences. We noted the specific body parts as accurately as possible and thus tried to map out a somatosensory 'homunculus'.

5.1.1.3. Somatosensory Evoked Potentials Recorded on Electrocorticography

In the first part of the study, we tried to map out cortical digit representations and therefore measured somatosensory evoked potentials (SEPs) on ECoG during the first 60 msec after shock stimulation of the median nerve, the ulnar nerve, and the individual digits in 4 patients. In the second part of the experiment, we studied cortical lip representation in relation to hand representation and recorded SEPs during the first 60 msec after shock stimulation of the median nerve, the ulnar nerve, and the lower lip in 3 patients.

Median and ulnar nerve were stimulated at the wrist. The digits were stimulated at the proximal interphalangeal joints using ring electrodes. The individual digits were separated with cotton balls to allow highly selective and exclusive stimulation. The lower lip was stimulated by two disk electrodes, the cathode was placed at the corner of the mouth, the anode paramedian. The stimulus was delivered by a Grass S88 stimulator (Grass Instrument Company, Quincy, MA) and consisted of monophasic, constant current 0.3 msec pulses. Stimulus frequency was set at 3.1 Hz. Stimulus intensity was 1.5 times motor threshold for stimulation of median and ulnar nerve, and twice sensory threshold for stimulation of the digits and the lower lip. Similar to the median nerve ECoG study (Section 4.1.1.3), ECoG was recorded simultaneously from a rectangular array of 48 chronically indwelling subdural grid electrodes with a diameter of 6 mm and 10 mm spacing center-to-center (PMT Corporation, Minneapolis, MN). ECoG was referenced to a scalp-EEG needle electrode at central vertex. Data were amplified (x 10,000) and filtered (bandpass 1 to 1,000 Hz) using Grass 12A5 amplifiers (Grass Instruments Company, Quincy, MA), digitized at a sampling rate of 4096 Hz (12 bits), and stored digitally for off-line data analysis. Two runs of 250 trials each were superimposed to assess reproducibility.

5.1.1.4. Correlation of Neuroelectric and Anatomical Data

Similar as already described for median nerve SEPs on ECoG (Section 4.1.1.4), source localizations were correlated with the results of cortical stimulations and anatomical structures obtained from intraoperative photographs, skull x-rays, and magnetic resonance images. Thus, also the exact location of central fissure in relation to the electrodes could be obtained. In the figures, source localizations were depicted as surface projections, i.e. as perpendicular projection of the source from the center of the sphere unto the cortical surface. Hand somatosensory cortex was spared from resection. Detailed neurological examinations were performed in all patients before and after surgery.

5.1.1.5. Data Analysis

As already described in Section 4.1.1.5, reproducibility of the data was tested by calculating run-to-run variability as the sum of squared differences between run #2 and run #1 divided by the sum of squares of run #1. Furthermore, we calculated the signal power to determine the peak latencies. Isopotential maps were computed at these peak latencies.

We applied single equivalent dipole models to study the three-dimensional location and time activities of the neuronal sources underlying the SEPs. To obtain the forward solution, we chose a homogeneous spherical model because measurements were made directly at the cortical surface avoiding inhomogenities. Furthermore, the surface of the brain approximates a sphere in the parieto-frontal region. Similar to Section 4.1.1.5, the radius of the sphere was determined from MRI scans, the conductivity of the brain was assumed as $(0.33 \; (\Omega m)^{-1})$ according to values reported in the literature [83, 136, 337], and the shape of the central sulcus was neglected in the model due to its minor effects on the electrical fields [80]. Thus, each dipole was uniquely given by 6 parameters – 3 location parameters representing its three-dimensional location, 2 orientation parameters representing its orientation, and one amplitude parameter representing its strength. The potential distribution generated by a dipole was calculated from these 6 parameters according to standard formulas [83, 84, 337, 338, 372] (see also Section 2.2). The goal of the modeling procedure thus was to determine which set of dipole parameters yielded a potential distribution that most closely reproduced the data. The dipole parameters were varied iteratively using the simplex algorithm [295] until the optimum combination of source location, orientation, and strength parameters was obtained. The dipole was allowed to move within the sphere to minimize the sum of squared differences between the model and the data. Variance accounted for was computed as the sum of squared differences between the model and the data divided by the sum of squares of the data across all recording positions.

Subsequently, we calculated the statistical means of the dipole location parameters over the entire time domain between the peak latencies to get an estimate of the center of activity during stimulation of median nerve, ulnar nerve, all the digits, and the lower lip.

5.1.2. Somatotopy as Studied on Scalp-EEG

5.1.2.1. Subjects and Procedures

We measured somatosensory evoked potentials (SEPs) on scalp-EEG during the first 60 msec after stimulation of the median nerve, the ulnar nerve, and the 5 digits in 3 normal subjects and one epilepsy patient evaluated for definitive localization of the seizure focus (total of 4 subjects). All subjects had given written informed consent to participate in the study.

The stimulus parameters were set identical to those already described in the previous section on ECoG (Section 5.1.1.3): 0.3 msec monophasic, constant current pulses; stimulation of median nerve and ulnar nerve at the wrist at 1.5 times motor threshold; stimulation of the digits at the proximal interphalangeal joints at double sensory threshold.

Scalp-EEG was recorded simultaneously from 32 (2 subjects) or 48 (2 subjects) gold disc electrodes (Grass Instruments Company, Quincy, MA). Reference electrode (linked ears), EEG recording matrix, amplification and filter settings were identical to the study of median nerve SEPs on scalp-EEG and have been described in detail in Section 4.1.2.1. Two runs of 500 trials each were superimposed to assess reproducibility. One run lasted for about 15 minutes, the whole experiment lasted for about 6-8 hours and was completed within one experimental session. The sequence of experimental conditions (stimulation sites) was varied randomly.

5.1.2.2. Correlation of Neuroelectric and Anatomical Data

In order to relate electrode positions to anatomical landmarks, we obtained skull x-rays in two planes (anterior-posterior and lateral views) with the electrodes in place, and defined a head coordinate system using a probe positioning system (Ptolhemus Navigation Sciences, McDonnel Douglas, Colchester, Vermont, U.S.A.). The anatomical reference points and the axes of the head coordinate system were set identical to those used for median nerve SEPs recorded on scalp-EEG and have been described in detail in Section 4.1.2.2.

5.1.2.3. Data Analysis

Like in the ECoG studies, reproducibility of the data was tested by calculating run-to-run variability. Signal power was calculated to determine the peak latencies. Two-dimensional isofield maps were computed at these peak latencies using a distance-weighted, least-squares approximation technique [241].

We applied source localization techniques to study the three-dimensional locations and time activities of the neuronal sources underlying the SEPs. Concerning the volume conductor, we assumed a 4-shell spherical model with concentric variations in conductivities for brain tissue, cerebrospinal fluid, skull, and scalp. The parameter values for the sphere radii and the conductivities were obtained in complete analogy to the procedure described for median nerve SEPs recorded on scalp-EEG (Section 4.1.2.3). Concerning the neuronal sources underlying SEPs, we assumed single equivalent dipoles which were uniquely determined by 6 parameters – 3 location parameters, 2 orientation parameters, and one amplitude parameter (see also Section 2.2). The goal of the modeling procedure was to determine which set of dipole parameters yielded a potential distribution that most closely reproduced the data. During the source localization procedure, the dipoles were allowed only to move within the innermost sphere, i.e. the brain, and it was tried to minimize the squared differences between the model and the data across all recording positions. Variance accounted for was computed for each dipole fit.

These source localization techniques were applied for each time slice in an interval between the peak latencies and for some short time intervals before the first and after the last peak latency. Source localizations were mapped onto the anatomical coordinates given by skull x-rays and the head coordinate system. The electro-anatomical correlations were shown as semi-schematic head figures with superimposed surface projections of the sources, i.e. perpendicular projections of the sources from the center of the sphere unto the scalp.

5.1.3. Somatotopy as Studied on Magnetoencephalography

5.1.3.1. Subjects and Procedures

We measured somatosensory evoked magnetic fields (SEFs) in 4 healthy volunteers during the first 60 msec after stimulation of the median nerve, the ulnar nerve, and the 5 digits of the left hand. All subjects had given written informed consent to participate in the study.

The stimulus parameters were set identical to those used for ECoG and scalp-EEG recordings and have been described in detail in Section 5.1.1.3: 0.3 msec monophasic, constant current pulses; stimulation of median nerve and ulnar nerve at the wrist at 1.5 times motor threshold; stimulation of the digits at the proximal interphalangeal joints at double sensory threshold.

Neuromagnetic recordings were performed using dual 7 channel magnetometers with dc-SQUIDs (second derivative gradiometer; coil diameter 18 mm; coil baseline 40 mm; Biomagnetic Technologies Inc., San Diego, CA, U.S.A.) in a large electromagnetically shielded room (Vacuumschmelze GmbH, Hanau, Germany). Magnetometer coils were located in a hexagonal array with 27 mm distance center-to-center in a dewar with a curved bottom. The overall noise level was 15 femtoTesla/Hz$^{1/2}$ (above 1Hz). It should be noted that the specifications of the magnetometer are slightly different compared to those for the median nerve SEF study presented in Section 4.1.3.1 because this study was performed at the Reed Neuromagnetism Laboratory, Department of Neurology, University of California, Los Angeles, U.S.A., whereas the somatotopy study was performed at the Neurological University Clinic, Vienna, Austria.

The recording strategy was similar to that used in the study on median nerve SEFs (Section 4.1.3.1). The magnetometer was placed at 9 different overlapping positions yielding a total of 63 and 45 distinct measurement points (Figs. 5.2.3.1, 5.2.3.2, and 5.2.3.3). The measurement matrix was centered at the C3/C4 position of the International 10-20-System. Data were filtered with a bandpass of 1 to 1,000 Hz, digitized at a sampling rate of 4000 Hz (12 bits), and stored digitally for off-line data analysis. Two runs of 500 trials each were superimposed to assess reproducibility.

The subjects were lying on a wooden non-magnetic bed with the head and upper body fixated with vacuum casts to prevent movement during the experiment. Furthermore, the stability of the position of the subject's head in relation to the magnetometer was determined by using the probe positioning system before and after each experimental session (see also Section 5.1.3.2).

As the main interest of our study was the relative location of cortical digit representations, each experimental session consisted of stimulation of median nerve, ulnar nerve, and all the digits without moving the dewar. One session lasted for approximately 1 to 1 1/2 hours. Usually, 2 to 3 sessions were performed on one day. Thus, a study on one subject could be completed in 3-5 experimental days. The sequence of probe positions was randomized though on subsequent days overlapping probe positions were used to test reproducibility of data over subsequent days.

5.1.3.2. Correlation of Neuromagnetic and Anatomical Data

In order to correlate neuromagnetic and anatomical data, a probe positioning system (Ptolhemus Navigation Sciences, McDonnel Douglas, Colchester, Vermont, USA) was used. Thus, a three-dimensional head coordinate system for each subject on the basis of three anatomical landmarks (nasion, both preauricular points) was constructed. This head coordinate system corresponded to that used for the scalp-EEG studies (see Sections 4.1.2.2 and 5.1.2.2). Furthermore, the probe positioning system enabled us to determine the position and orientation of the magnetometer in relation to the subject's head exactly. Thus, the correlation of neuromagnetic and anatomical data could be obtained in a more sophisticated and accurate way as compared to the study of median nerve SEFs (Section 4.1.3.2).

5.1.3.3. Data Analysis

Like in the ECoG and EEG studies, reproducibility of the data was tested by calculating run-to-run variability. We calculated the signal power to determine the peak latencies. Two-dimensional isofield maps were computed at these peak latencies using a distance-weighted, least-squares approximation technique [241].

We then applied source localization techniques to study the three-dimensional locations and time activities of the neuronal sources underlying the SEFs using single equivalent dipole models. Concerning the volume conductor, we assumed a homogenous spherical model as magnetic fields are not influenced by concentric changes in conductivity [314, 369] and spherical models are equally accurate as brain shaped models for source localization in the parieto-frontal region [157]. The radius of the sphere was calculated from the positions of the magnetometer as determined by the PPI which corresponded to the surface of the scalp. Concerning source configuration, we assumed single equivalent dipoles. Thus, each dipole was uniquely given by 5 parameters – 3 location parameters representing its three-dimensional location, one orientation parameter representing its orientation in a plane tangential to the surface of the sphere, and one amplitude parameter representing its strength. The field distribution generated by a dipole was calculated from these 5 parameters according to standard formulas [337, 370] (see also Section 2.2). The goal of the modeling procedure was to determine which set of dipole parameters yielded a field distribution that most closely reproduced the data. During the iterative search strategy the dipole was allowed to move only within a sphere approximating the human brain.

These source localization techniques were applied for each time slice in an interval between the peak latencies and for some short time intervals before the first and after the last peak latency. Source localizations for individual peaks as well as their statistical means were mapped onto the anatomical coordinates given by the head coordinate system. The magneto-anatomical correlations were displayed as semi-schematic head figures with superimposed surface projections of the sources, i.e. perpendicular projections of the sources from the center of the sphere unto the scalp.

5.2. Results

5.2.1. Somatotopy as Studied with Cortical Stimulations and Somatosensory Evoked Potentials on Electrocorticography

5.2.1.1. Cortical Stimulations

The results of cortical stimulations are shown in Figs. 5.2.1.1, 5.2.1.3, 5.2.1.7, and 5.2.1.9 (Patient #1), in Figs. 5.2.1.2, 5.2.1.8, and 5.2.1.10 (Patient #2), and in Figs. 5.2.1.11 (Patient #3). Skull x-rays with the subdural grid electrodes in place, and the stimulation results are shown. Black electrodes denote a positive motor response, gray electrodes a positive sensory response, and half black and gray electrodes a mixed motor/sensory response. Electrodes with negative stimulation effects (e.g. negative motor responses), with speech interference, and without stimulation effect were not differentiated and are shown as white.

The results were similar to those presented already in Section 4.2.1.1. Generally, anterior to central sulcus positive motor responses and posterior to central sulcus positive sensory responses could be obtained. However, this separation of primary motor and sensory cortex was not absolute as sometimes sensory responses were elicited anterior to central sulcus and motor responses posterior. Responses were strictly contralateral. The quality

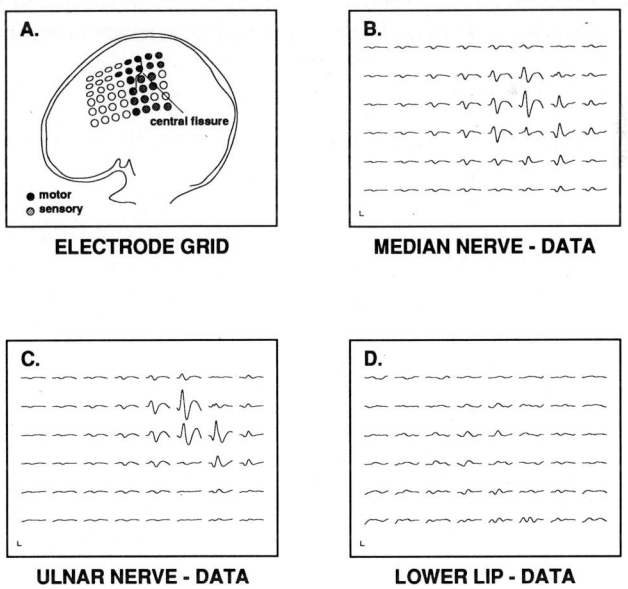

Fig. 5.2.1.1. Patient #1. **A.** Skull x-ray with subdural grid electrodes and results of cortical stimulations. Black electrodes denote motor responses, gray electrodes sensory responses, and half black and gray electrodes mixed motor/sensory responses. **B.** SEPs following stimulation of median nerve. **C.** SEPs following stimulation of ulnar nerve. **D.** SEPs following stimulation of the lower lips. Polarities: positive (+) up, negative (–) down. Calibration: horizontal = 10 msec, time scale begins at stimulus onset; vertical = 10 μV in (B) and (C), and 2 μV in (D)

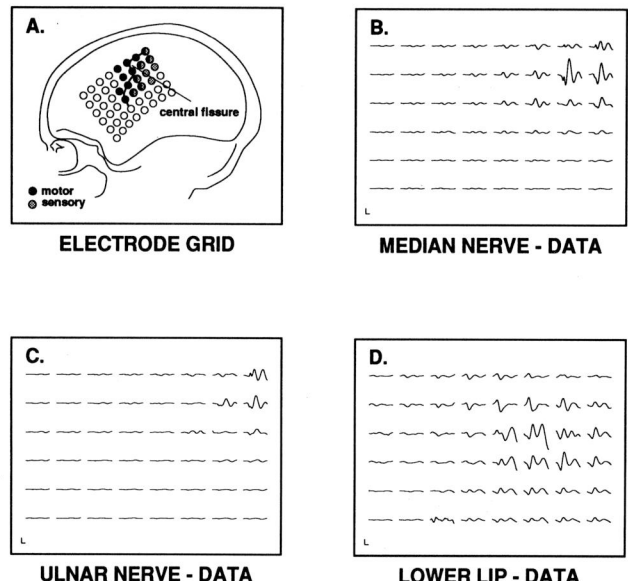

Fig. 5.2.1.2. Patient #2. **A.** Skull x-ray with subdural grid electrodes and results of cortical stimulations. Black electrodes denote motor responses, gray electrodes sensory responses, and half black and gray electrodes mixed motor/sensory responses. **B.** SEPs following stimulation of median nerve. **C.** SEPs following stimulation of ulnar nerve. **D.** SEPs following stimulation of the lower lips. Polarities: positive (+) up, negative (–) down. Calibration: horizontal = 10 msec, time scale begins at stimulus onset; vertical = 10 µV in (B) and (C), and 2 µV in (D)

of sensation induced by cortical stimulation was usually described as tingling or numbness.

Concerning somatotopy, stimulations of medial superior electrodes elicited sensory responses in arm and shoulder, stimulation of lateral inferior electrodes elicited sensory responses of the tongue, lips, and face. From electrodes in between, sensory responses in the hand could be obtained. However, cortical stimulations performed with this procedure did not allow a detailed somatotopic separation of the individual digits.

5.2.1.2. Somatosensory Evoked Potentials – Data

SEPs recorded on ECoG were reproducible within patients showing a run-to-run variability of 5-10%. Between patients there was some variability in waveforms and peak latencies which could be attributed to age differences and differences in lesion sites.

Median and ulnar SEPs (Figs. 5.2.1.1B, 5.2.1.1C, 5.2.1.2B, 5.2.1.2C, and 5.2.1.3) showed prominent peaks at latencies of about 20 and 30 msec with P20-N30 waveforms at the anterior recording electrodes and N20-P30 waveforms at the posterior recording electrodes. These components will be referred to as N20 and P30 in the following. In addition, positive peaks could be recorded at about 25 msec and negative peaks at about 35 msec. These components will be referred to as P25 and N35.

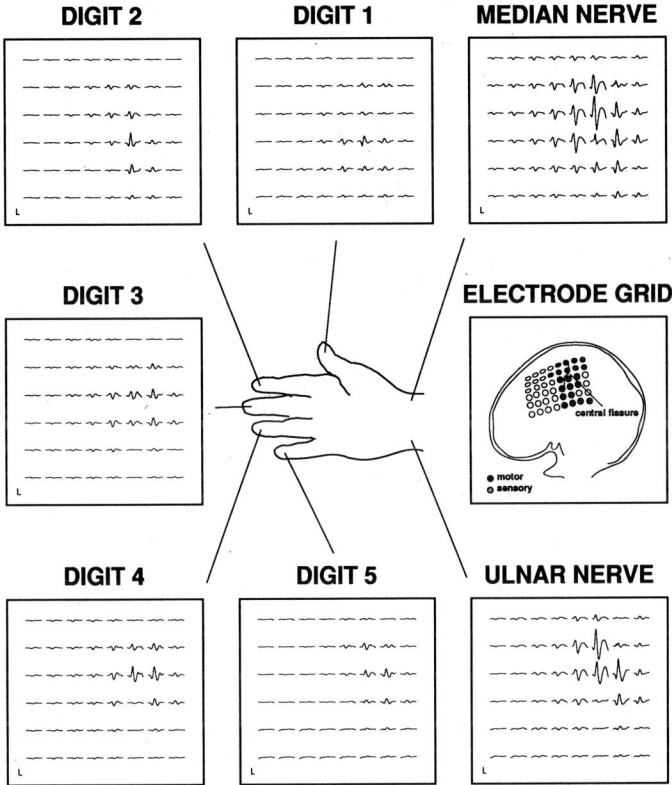

Fig. 5.2.1.3. Patient #1. Somatosensory evoked potentials following mixed nerve and digit stimulation. Skull x-ray with subdural grid electrodes and results of cortical stimulations. Black electrodes denote motor responses, gray electrodes sensory responses, and half black and gray electrodes mixed motor/sensory responses. Polarities: positive (+) up, negative (–) down. Calibration: horizontal = 10 msec, time scale begins at stimulus onset; vertical = 10 μV

Digit SEPs (Fig. 5.2.1.3) were of smaller amplitude compared to both median and ulnar nerve SEPs and peak latencies occurred slightly later at 23 and 33 msec. Otherwise digit SEPs exhibited similar qualitative features as median and ulnar nerve SEPs. Thus, at the anterior electrodes positive-negative waveforms (P23-N33 waveforms) could be recorded, and at the posterior electrodes negative-positive waveforms (N23-P33 waveforms). These components will be referred to as N23 and P33 components in the following. Furthermore, positive peaks could be recorded at about 28 msec and negative peaks at about 38 msec which will be termed P28 and N38 components, respectively.

Lower lip SEPs (Figs. 5.2.1.1D and 5.2.1.2D) showed prominent peaks at latencies of about 15 and 25 msec, with P15-N25 waveforms at the anterior electrodes and N15-P25 waveforms at the posterior electrodes, analogous to the N20 and P30 components for median and ulnar nerve. Positive P20 and negative N30 components also were seen, analogous to the P25 and N35 components for median and ulnar nerve.

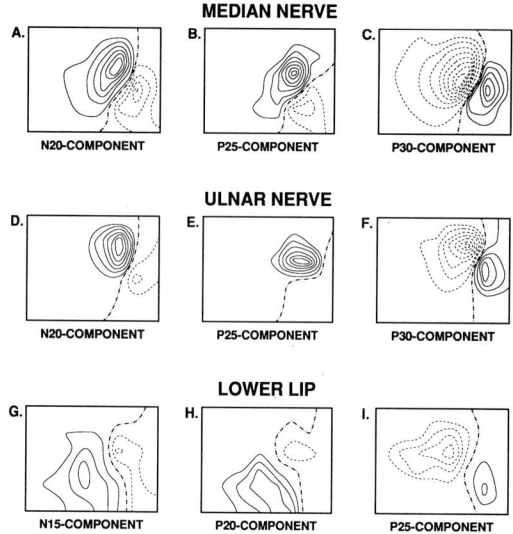

Fig. 5.2.1.4. Patient #1. Isopotential maps for median nerve, ulnar nerve, and lower lip SEPs. Median and ulnar nerve SEPs showed tangential N20 (A, D) and P30 components (C, F) with a phase reversal across central sulcus, and radial P25 components (B, E). Similar to hand SEPs, lower lip SEPs exhibited tangential N15 (G) and P25 components (I) with a phase reversal across central fissure. In contrast to hand SEPs, the lip P20 components (H) showed a primarily tangential pattern similar to that of the lip N15 component (G). The lip isopotential maps were lateral inferior to hand isopotential maps. Ulnar nerve maps were medial superior to median nerve maps. Isopotential maps: 10% isocontour lines; amplitudes scaled to maximum for each map; polarities: positive (+) solid, negative (–) dashed

5.2.1.3. Isopotential Maps for Median and Ulnar Nerve SEPs

Fig. 5.2.1.4 shows isopotential maps for N20, P25, and P30 components for median and ulnar nerve SEPs for Patient #1. N20 and P30 components showed tangential dipolar patterns with a phase reversal across central sulcus (Figs. 5.2.1.4A, 5.2.1.4C, 5.2.1.4D, and 5.2.1.4F). Isopotential maps for N20 and P30 were of opposite polarity, but otherwise were similar to each other concerning potential distribution. A line connecting the potential maxima showed an upward shift for ulnar nerve maps compared to those of median nerve. P25 for median and ulnar nerve stimulation showed more radial potential patterns with a single electropositivity (Fig. 5.2.1.4B and 5.2.1.4E). The potential maximum for median nerve stimulation was located lateral inferior compared to that of ulnar nerve stimulation. Fig. 5.2.1.5 shows the isopotential maps for the N20, P25, and P30 components for Patient #2, presented like for Patient #1. Despite some differences in the raw data (compare Figs. 5.2.1.1 and 5.2.1.2), the isopotential maps in this patient exhibited similar features as in Patient #1, namely tangential dipolar patterns for the N20 and P30 components (Figs. 5.2.1.5A, 5.2.1.5C, 5.2.1.5D, and 5.2.1.5F), radial dipolar patterns for the P25 components (Fig. 5.2.1.5B

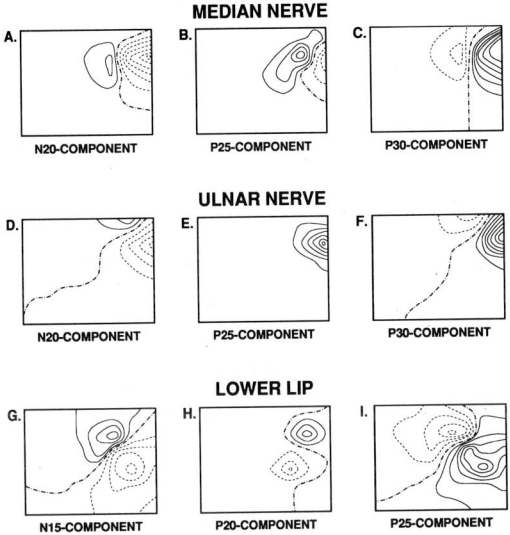

Fig. 5.2.1.5. Patient #2. Isopotential maps for median nerve, ulnar nerve, and lower lip SEPs. Median and ulnar nerve SEPs showed tangential N20 (A, D) and P30 components (C, F) with a phase reversal across central sulcus and radial P25 components (B, E). Similar to hand SEPs, lower lip SEPs exhibited tangential N15 (G) and P25 components (I) with a phase reversal across central fissure. Similar to Patient #1, the lip P20 component showed a primarily tangential pattern (H). In contrast to Patient #1, the lip P20 had the same polarity as the lip P25 component. The lip isopotential maps were lateral inferior to hand isopotential maps. Ulnar nerve maps were medial superior to median nerve maps. Isopotential maps: 10% isocontour lines; amplitudes scaled to maximum for each map; polarities: positive (+) solid, negative (−) dashed

and 5.2.1.5E), and an upward shift for the ulnar nerve maps in relation to the median nerve maps.

5.2.1.4. Isopotential Maps for Digit SEPs

Fig. 5.2.1.6 shows the isopotential maps for N23, P28, and P33 components for digit SEPs. Isopotential maps for N23 and P33 components during digit stimulation showed similar features as the isopotential maps for N20 and P30 components of median and ulnar nerve SEPs. N23 and P33 exhibited a tangential dipolar pattern with a phase reversal across central sulcus. N23 and P33 components were of opposite polarity, but otherwise were similar to each other concerning spatial potential distribution. A line connecting the potential maxima showed a progressive shift from lateral inferior to medial superior in the order thumb, index finger, middle finger, ring finger, and little finger. Like P25 components during median and ulnar nerve stimulation, P28 components during digit stimulation showed a radial dipolar pattern with a single electropositivity. Similar to N23 and P33 components, the P28 potential maxima showed a shift from lateral inferior to medial superior in the order thumb, index finger, middle finger, ring finger, and little finger.

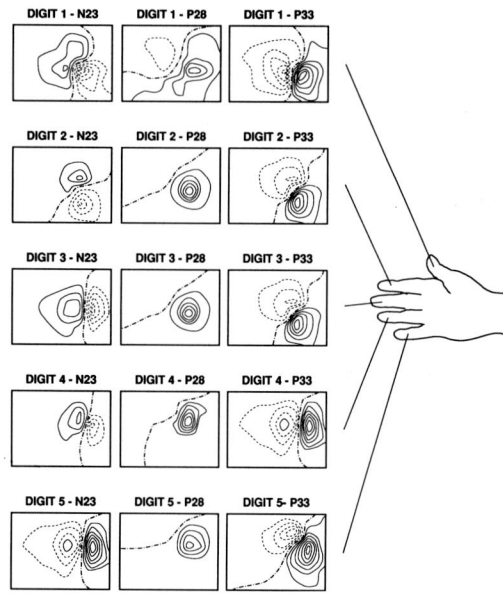

Fig. 5.2.1.6. Patient #1. Isopotential maps for digit SEPs. N23 and P33 components showed tangential dipolar patterns with a phase reversal across central sulcus. P28 components exhibited radial dipolar patterns with a single electropositivity. The maps showed a progressive shift from lateral inferior to medial superior in the order thumb, index finger, middle finger, ring finger, and little finger. Isopotential maps: 10% isocontour lines; amplitudes scaled to maximum for each map; polarities: positive (+) solid, negative (−) dashed

5.2.1.5. Isopotential Maps for Lip SEPs

Fig. 5.2.1.4 shows the isopotential maps for lip SEPs of Patient #1. Similar to the results of hand stimulation, the lip N15 and P25 maps also had tangential dipolar patterns with phase reversals across central sulcus (Figs. 5.2.1.4G and 5.2.1.4I). N15 and P25 isopotential maps were of opposite polarity. A line connecting the potential maxima of these components was located lateral inferior to the lines connecting the potential maxima during hand stimulation (Figs. 5.2.1.4A, 5.2.1.4C, 5.2.1.4D, 5.2.1.4F, 5.2.1.4G, and 5.2.1.4I). Compared to median and ulnar nerve isopotential maps, the lip N15 and P20 map peaks were more elongated. These results were reproducible across patients as can be seen from the isopotential maps for lip SEPs of Patient #2 (Figs. 5.2.1.5G and 5.2.1.5I).

In contrast to the qualitative similarities between the N20-P30 components for hand stimulation and the N15-P25 components for lip stimulation, the P20 components of lip stimulation showed marked differences compared to the P25 components of median and ulnar nerve stimulation. Whereas median and ulnar P25 components showed a primarily radial pattern, the lip P20 components showed a primarily tangential pattern similar to that of the lip N15 component in Patient #1 (Fig. 5.2.1.4H). In Patient #2, the lip P20 component also showed a primarily tangential pattern but was of opposite polarity compared to that in Patient #1 and thus was more similar to the lip P25 component (Fig. 5.2.1.5H).

5.2.1.6. Cortical Hand and Digit Representation

N20-P30 components of median and ulnar nerve SEPs as well as N23-P33 components for digit SEPs could be modeled by primarily tangentially oriented dipoles. Dipole orientations were opposite to each other at these latencies with the dipoles underlying the N20 and N23 components pointing anterior, and those underlying the P30 and P33 components pointing posterior. For the median and ulnar nerve P25-N35 components as well as for the digit P28-N38 components, we found primarily radially oriented sources which also were of opposite polarity at these latencies. The modeling procedure showed good fits to a single equivalent dipole (variance accounted for = $89.4 \pm 2.1\%$). The sources underlying the N20 and P30 components during median and ulnar nerve stimulation were located deeper compared to those underlying the P25 and N35 components (depth estimates below the cortical surface for the N20 and P30 components: mean = 7.4 mm, standard error = 0.3 mm (7.4 ± 0.3 mm); P25 and N35 components: 2.6 ± 0.3 mm; $p < 0.001$]. Similar, the sources underlying the N23 and P33 components during digit stimulation were located deeper compared to those underlying the P28 and N38 components (N23 and P33 components: 6.1 ± 1.2 mm; P25 and N35 components: 1.0 ± 0.8 mm; $p < 0.001$). There was no consistent difference between the sources of the N20-P30 components and of the P25-N35 components concerning the anterior-posterior or the medio-lateral axis. Similar results were obtained for digit P23-N33 and P28-N38 components. All locations of individual peaks and statistical means of the SEPs were located in postcentral gyrus within 11 mm of central sulcus. Surface projections of the source locations were next to electrodes from which sensory experiences in the hand could be elicited during cortical stimulations.

In all patients, median nerve sources were located lateral inferior to those of ulnar nerve. The mean distance between median and ulnar nerve cortex was 9.8 mm (range: 9-11 mm). Furthermore, cortical digit representations were arranged somatotopically from lateral inferior to medial superior in the order thumb, index finger, middle finger, ring finger, and little finger. Individual digits were well separated from each other. The mean distance between thumb and little finger was 16.1 mm (range: 15-18 mm). Whereas thumb, index finger, and middle finger were arranged around median nerve, ring finger and little finger were clustered around ulnar nerve.

In the following, we present the results of some selected subjects. In Fig. 5.2.1.7 somatotopy of hand somatosensory cortex is shown for Patient #1. Fig. 5.2.1.7A shows a skull x-ray with the grid recording matrix and Fig. 5.2.1.7B the results of source localizations and cortical stimulations. Statistical means of all dipole locations were calculated from 20 to 35 msec for median as well as for ulnar nerve, and from 23 to 33 msec for each digit.

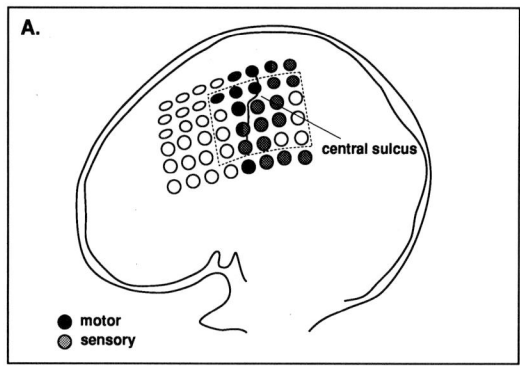

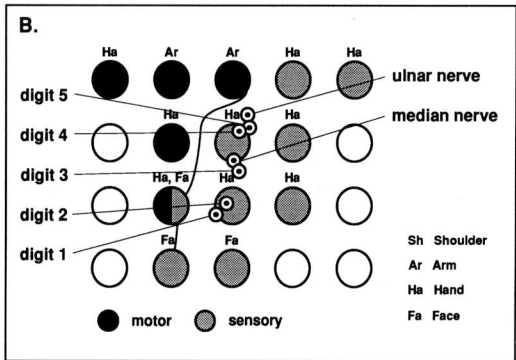

Fig. 5.2.1.7. Somatotopy of hand somatosensory cortex in Patient #1. **A.** Skull x-ray with subdural grid electrodes and results of cortical stimulations. Region inside dotted rectangle is enlarged for clarity at the bottom in (B). **B.** Source locations were calculated as the statistical means of all dipole locations from 20 to 35 msec for median as well as for ulnar nerve, and from 23 to 33 msec for each digit, and are presented as surface projections. The digits were arranged somatotopically from lateral inferior to medial superior in the sequence thumb, index finger, middle finger, ring finger, and little finger. Thumb, index finger, and middle finger were arranged around median nerve, and ring finger and little finger were clustered around ulnar nerve

These statistical means thus represent the center of cortical activity during peripheral nerve stimulation. Source locations are depicted as surface projections of these statistical means, i.e. radial projection of the three-dimensional intracerebral location of the source from the center of the sphere onto the cortical surface. All sources were within 8 mm of central fissure. The sources were well separated from each other and showed a somatotopic organization with the sources during ulnar nerve stimulation located medial superior and those during median nerve stimulation lateral inferior. The distance between median and ulnar nerve cortical representation was 9 mm in this patient. The sources underlying digit stimulation were arranged in a somatotopic way from lateral inferior to medial superior in the sequence thumb, index finger, middle finger, ring finger, and little finger. The distance between cortical thumb and little finger

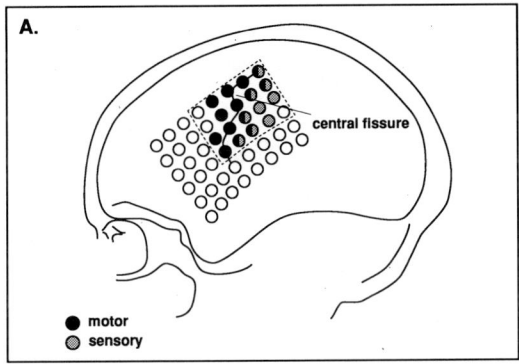

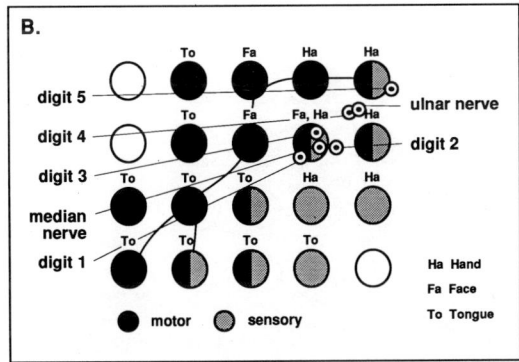

Fig. 5.2.1.8. Somatotopy of hand somatosensory cortex in Patient #2. **A.** Skull x-ray with subdural grid electrodes and results of cortical stimulations. Region inside dotted rectangle is enlarged for clarity at the bottom in (B). **B.** Source locations were calculated as the statistical means of all dipole locations from 20 to 35 msec for median as well as for ulnar nerve, and from 23 to 33 msec for each digit, and are presented as surface projections. The digits were arranged somatotopically from lateral inferior to medial superior in the sequence thumb, index finger, middle finger, ring finger, and little finger. Thumb, index finger, and middle finger were arranged around median nerve, and ring finger and little finger were clustered around ulnar nerve

representation was 15 mm in this patient. Thumb, index finger, and middle finger were arranged around median nerve cortical representation, whereas ring finger and little finger were clustered around ulnar nerve cortex. Furthermore, the results of source localizations agreed with the results of cortical stimulations as all sources were near electrodes where cortical stimulations elicited sensory responses in the hand.

Fig. 5.2.1.8 shows somatotopy of cortical digit representation for Patient #2, presented similar as for Patient #1. The results concerning the somatotopic arrangement of the median nerve, the ulnar nerve, and the individual digits for this patient were analogous to those for Patient #1. In this patient, all sources were within 11 mm of central fissure. The distance between median and ulnar nerve cortical representation was 10 mm, the distance between thumb and little finger representation was 18 mm.

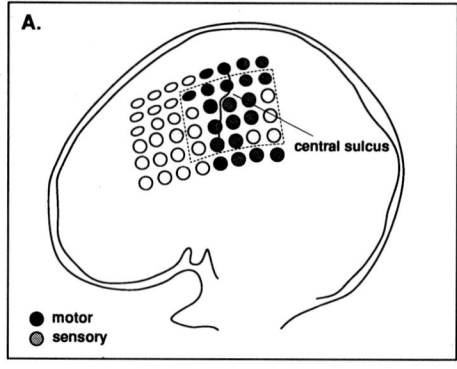

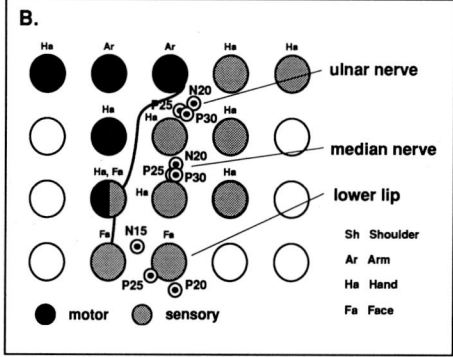

Fig. 5.2.1.9. Somatotopy of lip somatosensory cortex in relation to hand somatosensory cortex in Patient #1. **A.** Skull x-ray with subdural grid electrodes and results of cortical stimulations. Region inside dotted rectangle is enlarged for clarity at the bottom in (B). **B.** Dipole locations at the individual peak latencies are shown as surface projections. The sources underlying stimulation of different peripheral fields were well separated from each other and showed a somatotopic organization with the sources during ulnar nerve stimulation located medial superior, those during stimulation of the lower lips lateral inferior, and those during stimulation of the median nerve in between

5.2.1.7. Cortical Lip Representation in Relation to Hand Representation

In contrast to median nerve, ulnar nerve, and digit SEPs, all components of lip SEPs could be modeled by primarily tangential dipoles. Similar to hand SEPs, dipoles reversed orientations for the N15 and P25 components. Whereas N15 could be modeled by a tangential dipole pointing in the anterior direction, P25 could be modeled by a tangential dipole of reversed orientation, i.e. pointing in the posterior direction. There was no consistent difference concerning depth between the lip N15 and P25 sources on the one hand, and the lip P20 and N30 sources on the other hand. However, the lip SEP sources were deeper (9.2 ± 1.2 mm) than either median nerve SEP (4.8 ± 0.8 mm) or ulnar nerve SEP sources (4.8 ± 0.9 mm; $p < 0.001$) (note that these values represent pooled depth estimates for N20, P25, and P30). The dipole fits for lip SEPs were worse with more

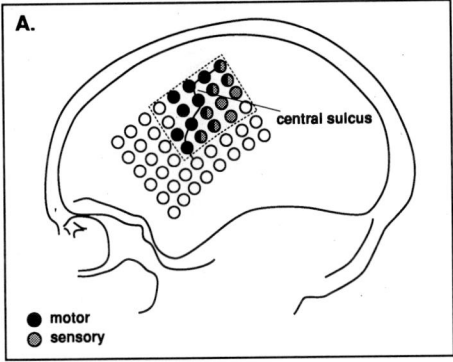

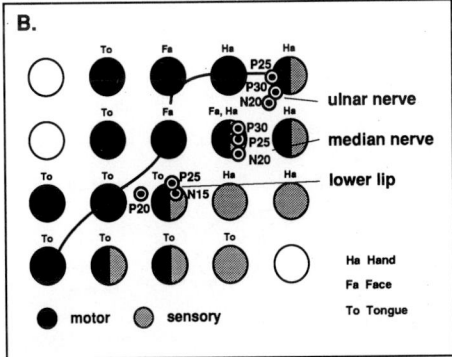

Fig. 5.2.1.10. Somatotopy of lip somatosensory cortex in relation to hand somatosensory cortex in Patient #2. **A.** Skull x-ray with subdural grid electrodes and results of cortical stimulations. Region inside dotted rectangle is enlarged for clarity at the bottom in (B). **B.** Dipole locations at the individual peak latencies are shown as surface projections. The sources underlying stimulation of different peripheral fields were well separated from each other and showed a somatotopic organization with the sources during ulnar nerve stimulation located medial superior, those during stimulation of the lower lips lateral inferior, and those during stimulation of the median nerve in between

variability (65.0 ± 4.6 %) than those of either median SEPs (85.5 ± 2.6 %) or ulnar SEPs (91.0 ± 1.7 %) (p < 0.001).

Somatotopy of cortical lip representation in relation to cortical hand representation is shown for all three patients in Figs. 5.2.1.9, 5.2.1.10, and 5.2.1.11. The dipole locations at the individual peak latencies and the results of cortical stimulations are depicted in these figures. Similar to the results on digit SEPs, source locations are depicted as surface projections, i.e. radial projection of the three-dimensional intracerebral location of the source from the center of the sphere onto the cortical surface. All sources were located in postcentral gyrus within 11 mm of central fissure. The sources underlying stimulation of different peripheral fields were well separated from each other and showed a somatotopic organization with the sources during ulnar nerve stimulation located medial superior, those during stimulation of the lower lips lateral inferior, and those during stimulation of

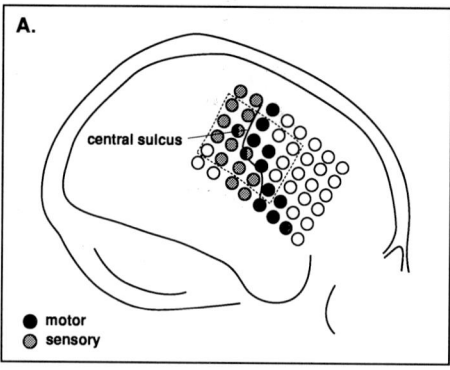

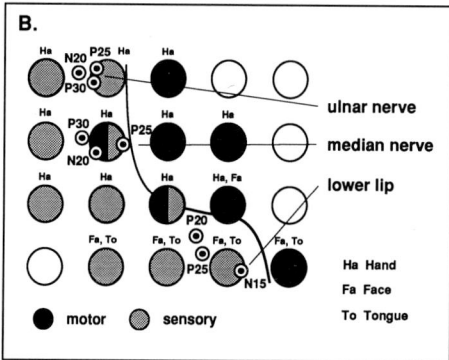

Fig. 5.2.1.11. Somatotopy of lip somatosensory cortex in relation to hand somatosensory cortex in Patient #3. **A.** Skull x-ray with subdural grid electrodes and results of cortical stimulations. Region inside dotted rectangle is enlarged for clarity at the bottom in (B). **B.** Dipole locations at the individual peak latencies are shown as surface projections. The sources underlying stimulation of different peripheral fields were well separated from each other and showed a somatotopic organization with the sources during ulnar nerve stimulation located medial superior, those during stimulation of the lower lips lateral inferior, and those during stimulation of the median nerve in between

the median nerve in between. Source localizations obtained at different latencies during stimulation of the same peripheral receptive field (N20, P25, and P30 for median and ulnar nerve, as well as N15, P20, and P25 for lower lip) were close to each other and showed no consistent differences concerning the anterior-posterior or the medio-lateral direction. Source localizations showed good agreement with the results of cortical stimulations as the sources underlying median and ulnar nerve SEPs were near electrodes where cortical stimulations elicited sensory responses in the hand and the sources of lip SEPs were near electrodes where cortical stimulations produced lip or face sensations.

5.2.2. Somatotopy as Studied on Scalp-EEG

5.2.2.1. Data

SEPs recorded on scalp-EEG were reproducible within subjects showing a run-to-run variability of 5-8%. Furthermore, data were similar across subjects. Figs. 5.2.2.1 and 5.2.2.2 show the data for Subjects #1 and 2, respectively.

Median and ulnar nerve SEPs showed prominent peaks at latencies of about 20 (mean±s.e. = 20.1±0.3) msec and 30 (29.8±0.4) msec with positive-negative waveforms at the frontal electrodes and negative-positive waveforms at the parietal electrodes. According to their peak latencies and polarities, these waveforms have been termed P20-N30 components (frontal electrodes) and N20-P30 components (parietal electrodes) [4, 7, 117]. For sake of simplicity, we will refer to these components as N20 and P30. In addition, positive peaks were recorded at about 25 (24.8±0.3) msec and negative peaks at about 35 (35.1±0.6) msec. These components could be seen best at the central electrodes. According to the nomenclature mentioned above [4, 7, 117], we will refer to these components as P25 and N35. Ulnar nerve SEPs were of slightly smaller amplitude compared to median nerve SEPs. Peak latencies for median and ulnar nerve did not show significant differences.

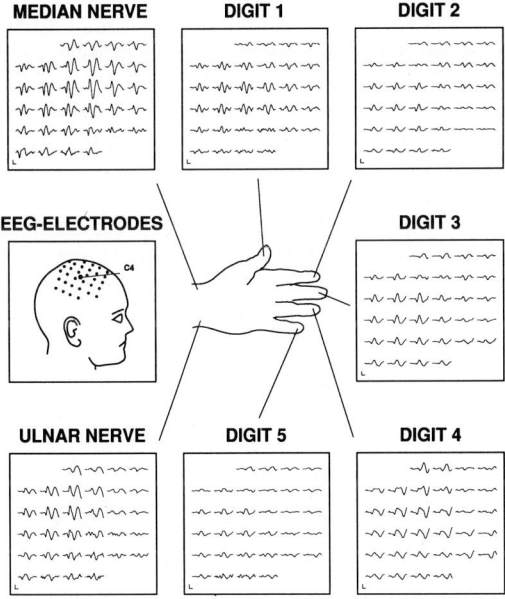

Fig. 5.2.2.1. Scalp-EEG recording matrix and data for Subject #1. Scalp-EEG was recorded simultaneously from a rectangular array of 32 electrodes which were arranged in a 6 times 6 matrix centered at the C4 position of the International 10-20-System (2 electrodes each at the posterior superior and anterior inferior corners of this grid were omitted). The electrodes were spaced 3 cm center-to-center. Median nerve, ulnar nerve, and the 5 digits of the left hand were stimulated. Calibrations: horizontal = 10 ms, time scale begins with stimulus onset; vertical = 1 μV

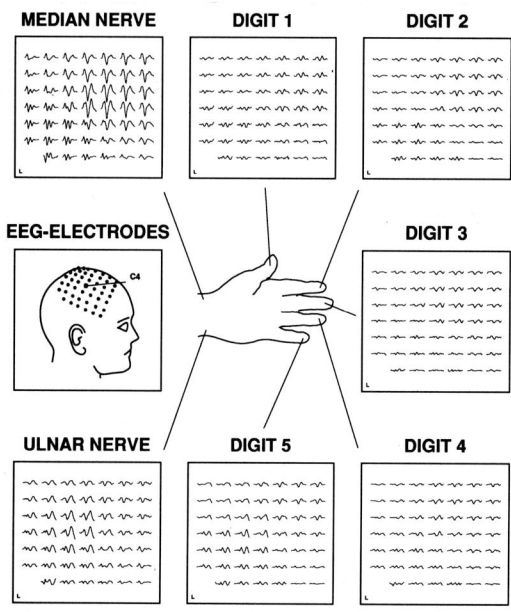

Fig. 5.2.2.2. Scalp-EEG recording matrix and data for Subject #2. Scalp-EEG was recorded simultaneously from a rectangular array of 48 electrodes which were arranged in a 7 times 7 matrix centered at the C4 position of the International 10-20-System (1 electrode at the posterior superior corner of this grid was omitted). The electrodes were spaced 3 cm center-to-center. Median nerve, ulnar nerve, and the 5 digits of the left hand were stimulated. Calibrations: horizontal = 10 ms, time scale begins with stimulus onset; vertical = 1 μV

Digit SEPs were of smaller amplitude compared to both median and ulnar nerve SEPs, but exhibited similar qualitative features. Peak latencies occurred at 23 (22.7 ± 0.4) msec and 33 (32.5 ± 0.5) msec. Similar to median and ulnar nerve SEPs, at the frontal electrodes positive-negative P23-N33 waveforms could be recorded, and at the parietal electrodes negative-positive N23-P33 waveforms were recordable. For sake of simplicity, we will refer to these waveforms as N23 and P33 components in the following. Furthermore, positive peaks were recorded at about 28 (27.6 ± 0.5) msec and negative peaks at about 38 (38.1 ± 0.6) msec which will be termed P28 and N38 components. Similar to mixed nerve stimulation, these components could be best identified at the central electrodes.

5.2.2.2. Isopotential Maps for Median and Ulnar Nerve SEPs

Fig. 5.2.2.3 shows isopotential maps for N20, P25, and P30 components for median and ulnar nerve SEPs for Subject #1. N20 and P30 components showed tangential dipolar patterns with a phase reversal between the frontal and parietal electrodes (Figs. 5.2.2.3A, 5.2.2.3C, 5.2.2.3D, and 5.2.2.3F). Phase reversal lines were slightly oblique from inferior anterior to superior posterior and thus followed approximately central sulcus. Isopotential maps for N20 and P30 were of opposite polarity, but otherwise were similar to each

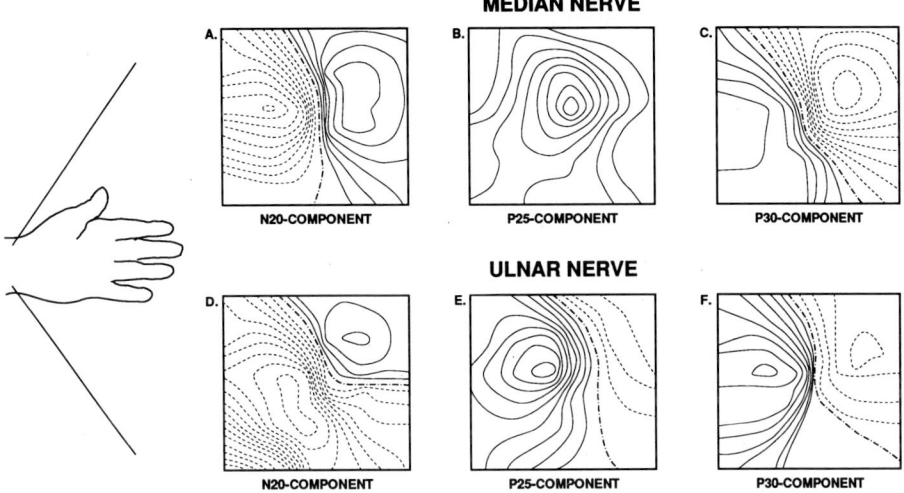

MEDIAN NERVE

A.   N20-COMPONENT
B.   P25-COMPONENT
C.   P30-COMPONENT

ULNAR NERVE

D.   N20-COMPONENT
E.   P25-COMPONENT
F.   P30-COMPONENT

Fig. 5.2.2.3. Subject #1. Isopotential maps for N20, P25, and P30 components during median and ulnar nerve stimulation. N20 (A, D) and P30 (C, F) components showed tangential dipolar patterns with a phase reversal between the frontal and parietal electrodes following approximately central sulcus. Isopotential maps for N20 and P30 were of opposite polarity, but otherwise were similar to each other concerning the spatial potential distribution. P25 components showed radial potential patterns with a single electropositivity (B, E). Median nerve maps were lateral inferior to ulnar nerve maps (compare A-C and D-F). Isopotential maps: 10% isocontour lines; each map scaled to maximum; polarities: positive (+) solid, negative (−) dashed

other concerning spatial potential distribution. A line connecting the potential maxima showed an upward shift for ulnar nerve maps compared to those of median nerve. The P25 components underlying median and ulnar nerve stimulation showed radial potential patterns with a single electropositivity (Figs. 5.2.2.3B and 5.2.2.3E). The potential maximum for median nerve stimulation was located lateral inferior to that of ulnar nerve stimulation.

In Fig. 5.2.2.4, isopotential maps for Subject #2 are presented. The median and ulnar nerve N20 components showed similar features as already described for Subject #1. These components showed a tangential dipolar pattern with a phase reversal between the frontal and parietal recording positions. The phase reversal lines were slightly oblique approximately following central fissure. Furthermore, the ulnar nerve N20 map was located superior medial to the median N20 map. These map features also were reproducible across the other subjects. Furthermore, these results were qualitatively similar to those obtained from intracranial recordings on ECoG (see Section 5.2.1.3).

5.2.2.3. Isopotential Maps for Digit SEPs

Fig. 5.2.2.5 shows the isopotential maps for N23, P28, and P33 components for digit SEPs for Subject #1. Isopotential maps for N23 and

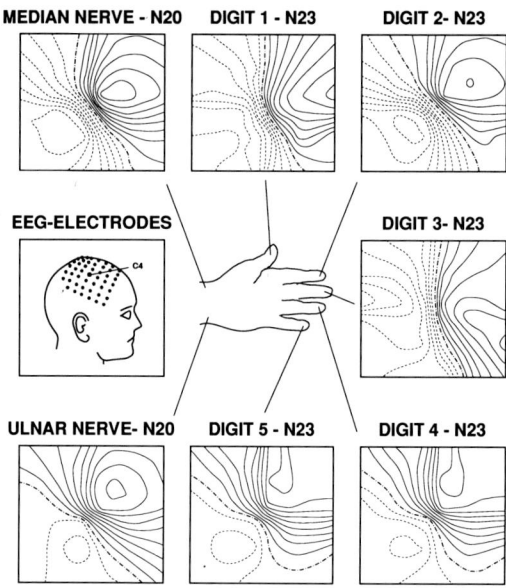

Fig. 5.2.2.4. Subject #2. Isopotential maps for N20 (median and ulnar nerve stimulation) and for N23 components (digit stimulation). N20 and N23 components showed tangential dipolar patterns with a phase reversal between the frontal and parietal recording positions approximately following central fissure. The median nerve N20 map was located lateral inferior to the ulnar nerve N20 map. The thumb N23 map was lateral inferior to the little finger N23 map. Furthermore, the maps of index finger, middle finger, and ring finger showed a slight shift from lateral inferior to medial superior which, however, was less evident from visual inspection of the maps. Isopotential maps: 10% isocontour lines; each map scaled to maximum; polarities: positive (+) solid, negative (−) dashed

P33 components during digit stimulation showed similar features as the isopotential maps for N20 and P30 of median and ulnar nerve SEPs, namely a tangential dipolar pattern with a phase reversal between the frontal and parietal electrodes (compare Figs. 5.2.2.3 and 5.2.2.5). For each individual digit, N23 and P33 maps were of opposite polarity, but otherwise were similar concerning the spatial potential distribution. The phase reversal lines were oriented slightly oblique from inferior anterior to superior posterior thus approximately following central sulcus. The P28 components showed primarily radial patterns, similar to the P25 components during mixed nerve stimulation. Concerning somatotopy, the isopotential maps showed a shift from lateral inferior to medial superior for thumb versus little finger. This can be evaluated best by connecting the potential maxima for the N23 and P33 components or by inspection of the potential maxima for the P28 components. Also the maps of the digits in between, i.e. of index finger, middle finger, and ring finger, showed a slight shift according to the sensory sequence, i.e from lateral inferior to medial superior. However, this somatotopic shift was less evident from visual inspection of the maps compared to the differences between thumb and little finger maps.

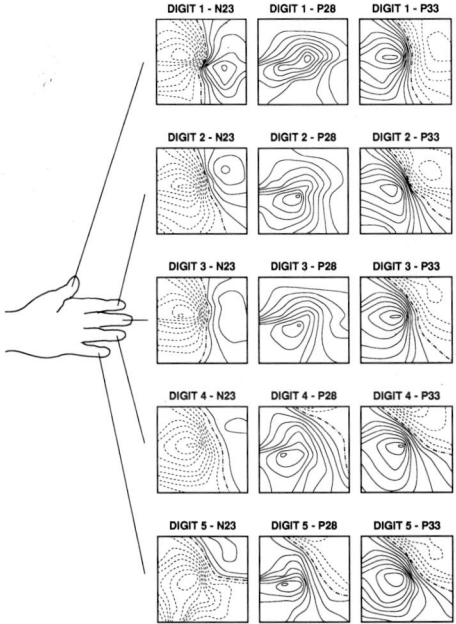

Fig. 5.2.2.5. Subject #1. Isopotential maps for N23, P28, and P33 components during digit stimulation. N23 and P33 components showed tangential dipolar patterns with a phase reversal between the frontal and parietal electrodes following approximately central sulcus. Isopotential maps for N23 and P33 components were of opposite polarity, but otherwise were similar to each other concerning the spatial potential distribution. P28 components showed radial potential patterns with single electropositivities. Concerning somatotopy, the isopotential maps showed a shift from lateral inferior to medial superior for thumb versus little finger. Furthermore, the maps of index finger, middle finger, and ring finger showed a slight shift from lateral inferior to medial superior which, however, was less evident from visual inspection of the maps. Isopotential maps: 10% isocontour lines; each map scaled to maximum; polarities: positive (+) solid, negative (−) dashed

These findings were reproducible across subjects. Fig. 5.2.2.4 shows the digit N23 maps for Subject #2 which exhibited similar features. These maps showed a tangential dipolar pattern with phase reversals approximately following central fissure and a shift in the medial superior direction for little finger in comparison to thumb. These findings were similar to the results of digit SEPs on electrocorticography though the maps showed a more widespread pattern on scalp-EEG (see Section 5.2.1.4).

5.2.2.4. Cortical Hand and Digit Representation

N20 and P30 components (median and ulnar nerve SEPs) as well as N23 and P33 components (digit SEPs) could be modeled by primarily tangentially oriented dipoles. Dipole orientations were opposite to each other at these latencies with the dipoles underlying the N20 and N23 components pointing anterior and slightly upwards, and those underlying the P30 and P33 components pointing posterior and slightly downwards. The dipoles thus were oriented approximately perpendicular to central fissure.

Furthermore, we found radial dipoles for the P25 and N35 components (median and ulnar nerve SEPs) as well as for the P28 and N38 components

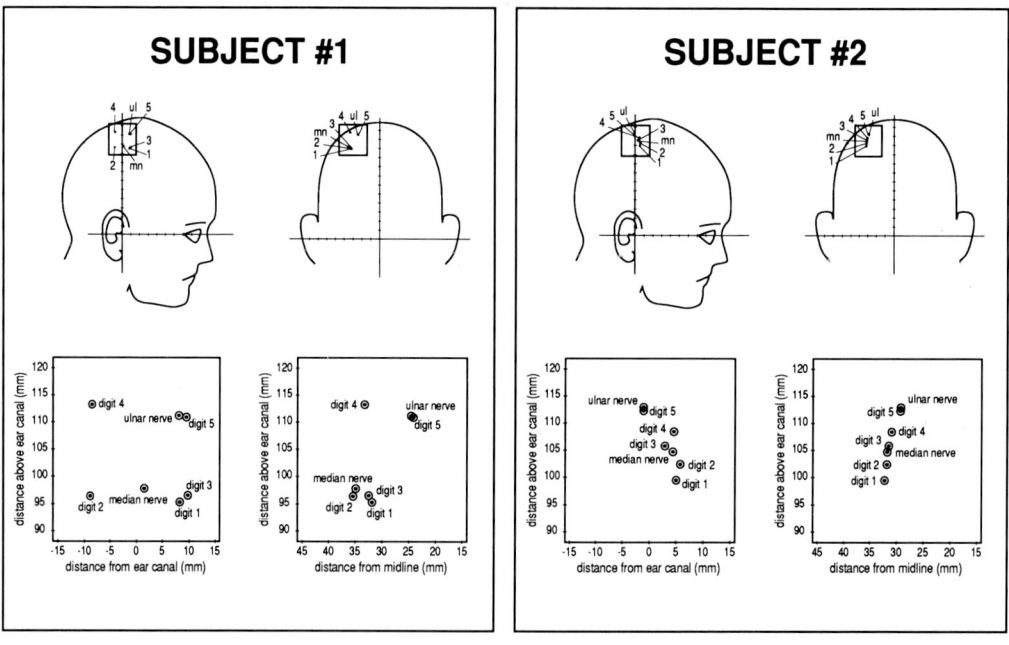

Fig. 5.2.2.6. Source locations for Subjects #1 and #2. Source locations for the N20 components (median and ulnar nerve stimulation) as well as for the N23 components (digit stimulation) are depicted as circled dots in sagittal and coronal schematic views. Regions inside boxes are enlarged at the bottom for clarity. In Subject #1, there was a good separation of digits 1-3 and median nerve on the one hand, and digits 4-5 and ulnar nerve on the other hand. Some adjacent digits showed overlap or reversal in the sensory sequence (cf. digits 2 and 3; digits 4 and 5). In Subject #2, cortical digit representation followed the sensory sequence from lateral inferior to medial superior in the order thumb, index finger, middle finger, ring finger, and little finger. Furthermore, there was a good separation of digits 1-3 and median nerve versus digits 4-5 and ulnar nerve

(digit SEPs). Concerning orientation P25 and P28 dipoles were pointing towards the surface of the brain, whereas N35 and N38 dipoles were directed towards the center of the sphere. The variance accounted for by the model averaged 88.7 ± 4.3%. The dipole locations during stimulation of a specific nerve or digit were very close to each other at successive peak latencies. There were no significant differences concerning depth between the tangential sources (N20 and P30 for mixed nerve stimulation as well as N23 and P33 for digit stimulation) on the one hand, and the radial sources (mixed nerve P25 and N35 as well as digit P28 and N38) on the other hand.

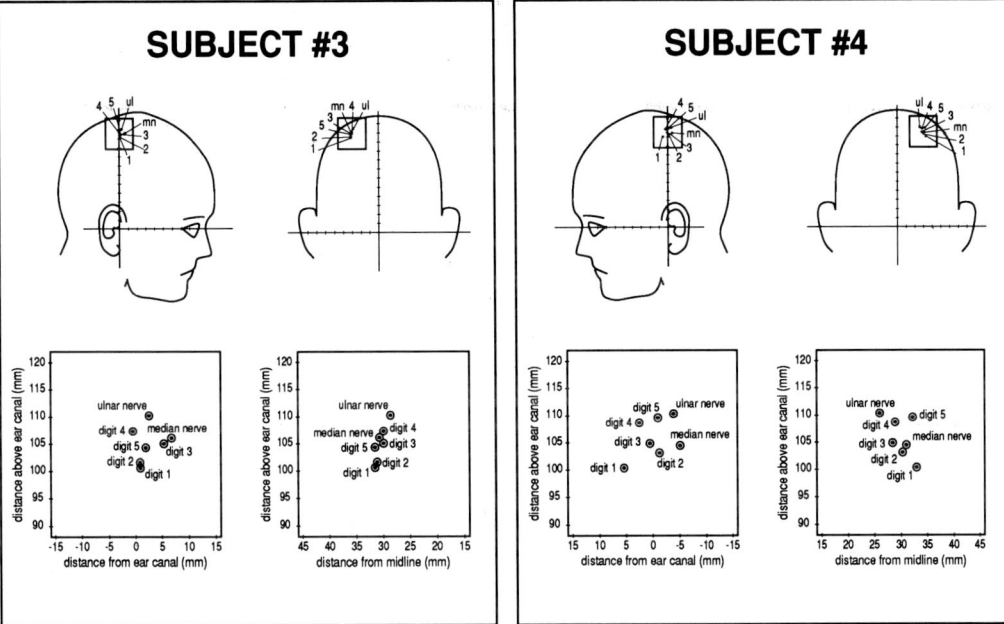

Fig. 5.2.2.7. Source locations for Subjects #3 and #4. Source locations for the N20 components (median and ulnar nerve stimulation) as well as for the N23 components (digit stimulation) are depicted as circled dots in sagittal and coronal schematic views. Regions inside boxes are enlarged at the bottom for clarity. In Subject #3, cortical digit representations were closer to each other compared to the other subjects. Whereas digits 1-4 followed the sensory sequence, digit 5 was located close to digit 3. Digit 1-3 and digit 5 were arranged around median nerve, and digit 4 was close to ulnar nerve. In Subject #4, there was a somatotopic arrangement of the digits following the sensory sequence and a separation of digits 1-3 and median nerve versus digits 4-5 and ulnar nerve

Concerning somatotopy, median nerve sources were located lateral inferior to ulnar nerve sources. The distance between these source averaged 10.8 mm (range: 7-18 mm). Cortical thumb representation was located lateral inferior to little finger representation with an average distance of 12.5 mm (range: 4-17 mm) between these cortical regions. Concerning the arrangement of the digits in between, i.e. index finger, middle finger, and ring finger, we got variable results. In two subjects these digits followed exactly the sensory sequence, whereas in the other two subjects some adjacent digits showed overlapping or reversed cortical representations. Generally, thumb, index finger, and middle finger were clustered around median nerve cortical representation, whereas ring finger and little finger were arranged close to ulnar nerve cortical representation.

In the following, we present the results of the individual subjects. The source locations underlying the N20 components (mixed nerve stimulation)

and the N23 components (digit stimulation) are depicted in sagittal and coronal schematic views. We chose the N20 and N23 peaks, respectively, as these components reflect the first cortical activity and thus less overlap from other cortical sources as for the subsequent components should occur. In Subject #1 (Fig. 5.2.2.6, left), we found a good separation of digits 1-3 and median nerve on the one hand, and digits 4-5 and ulnar nerve on the other hand. Some adjacent digits showed overlap or reversal in the sensory sequence (cf. digits 2 and 3; digits 4 and 5). In Subject #2 (Fig. 5.2.2.6, right), cortical digit representation followed exactly the sensory sequence from lateral inferior to medial superior in the order thumb, index finger, middle finger, ring finger, and little finger. Furthermore, there was a good separation of digits 1-3 and median nerve versus digits 4-5 and ulnar nerve. In Subject #3 (Fig. 5.2.2.7, left), cortical digit representations were closer to each other compared to the other subjects. Whereas digits 1-4 followed the sensory sequence, digit 5 was located close to digit 3. Digits 1-3 and digit 5 were arranged around median nerve, and digit 4 was close to ulnar nerve. For Subject #4 (Fig. 5.2.2.7), finally, we found a somatotopic arrangement of the digits following the sensory sequence and a separation of digits 1-3 and median nerve versus digits 4-5 and ulnar nerve similar to the results in Subject #2.

5.2.3. Somatotopy as Studied on Magnetoencephalography

5.2.3.1. Data

The data were reproducible within subjects showing a run-to-run variability of 6-9% and were similar across subjects. The data and the MEG recording matrix for Subject #1 are shown in Fig. 5.2.3.1. Median and ulnar nerve somatosensory evoked magnetic field (SEFs) showed prominent peaks at about 20 (mean $\pm$ s.e. = 19.6 ± 0.5) msec and 30 (28.7 ± 0.4) msec. According to their average peak latencies, these waveforms will be referred to as SEF_{20} and SEF_{30}. Whereas peak amplitudes for median nerve SEFs were larger compared to those of ulnar nerve, peak latencies for median and ulnar nerve SEFs showed no significant differences. At the superior recording positions, there was a sequence of positive SEF_{20} waveforms (corresponding to a magnetic field emerging from the head) and negative SEF_{30} waveforms (corresponding to a magnetic field re-entering the head). On the contrary, at the inferior recording positions a sequence of negative SEF_{20} and positive SEF_{30} waveforms could be measured.

Digit SEFs showed longer peak latencies and smaller amplitudes, but otherwise were qualitatively similar to mixed nerve SEFs (Fig. 5.2.3.1). Peak latencies for digit SEFs occurred at 23 (22.3 ± 0.3) msec and 33 (32.5 ± 0.4) msec and therefore will be termed SEF_{23} and SEF_{33}. Similar to mixed nerve stimulation, at the superior recording positions positive-negative SEF_{23}-SEF_{33} wave sequences were seen, and at the inferior recording positions negative-positive SEF_{23}-SEF_{33} wave sequences could be recorded.

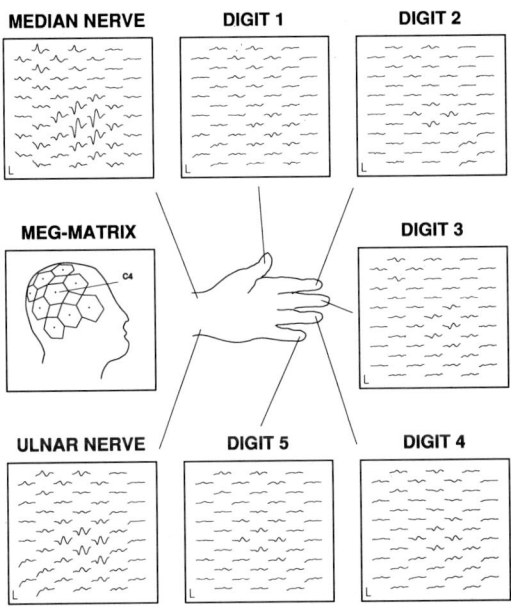

Fig. 5.2.3.1. MEG recording matrix and data for Subject #1. The 7-channel magnetometer was placed at 9 different overlapping positions yielding 45 distinct measurement points. The measurement matrix was centered at the C4 position of the International 10-20-System. Median nerve, ulnar nerve, and the 5 digits of the left hand were stimulated. Calibrations: horizontal = 10 ms, time scale begins with stimulus onset; vertical = 100 fT

5.2.3.2. Isofield Maps for Median, Ulnar Nerve and Digit SEFs

Fig. 5.2.3.2 shows the isofield maps for the median and ulnar nerve SEF_{20} as well as for the digit SEF_{23} components for Subject #1. Mixed nerve and digit SEF maps were qualitatively similar. The magnetic field pattern showed a horizontal reversal of field direction between the upper and lower recording positions, and emerged from the head medial superior and re-entered the head lateral inferior. The lines of magnetic field reversal were slightly oblique from anterior superior to posterior inferior and thus were approximately perpendicular to central sulcus. These magnetic field patterns are therefore consistent with tangential dipolar current sources located in central sulcus and pointing in the anterior direction. A shift in the medial superior direction could be observed for the ulnar nerve map in comparison to the median nerve map. Similarly, the isofield map for the little finger was located medial superior to the thumb isofield map. For the digits in between, i.e. index finger, middle finger, and ring finger, there was a slight shift of the maps in the medial superior direction which, however, was difficult to discern from visual inspection of map features.

Fig. 5.2.3.3 shows the isofield maps for the SEF_{23} and SEF_{33} components following digit stimulation for Subject #2. Similar as in Subject #1, isofield maps for the SEF_{23} reversed in direction between the upper and lower recording positions with the magnetic field emerging from the head medial

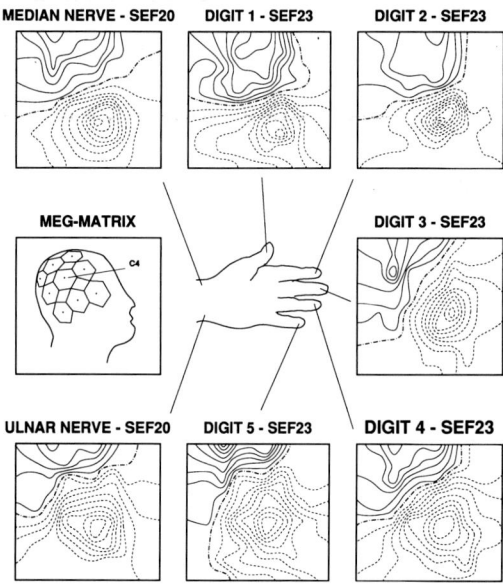

Fig. 5.2.3.2. Isofield maps for the SEF_{20} components (mixed nerve stimulation) and for the SEF_{23} components (digit stimulation) in Subject #1. The magnetic field patterns showed a horizontal reversal of field directions between the upper and lower recording positions emerging from the head medial superior and re-entering the head lateral inferior. Reversals of field direction showed a shift from lateral inferior to medial superior for digits 1 to 5, and for ulnar nerve compared to median nerve. Isofield maps: 10% isocontour lines; each map scaled to maximum; solid lines: magnetic field emerging from the head; dashed lines: magnetic field re-entering the head

superior and re-entering the head lateral inferior. Isofield maps for the SEF_{33} showed similar magnetic field patterns compared to those of SEF_{23}, but were inverted in direction. For these components, the magnetic field emerged from the head lateral inferior and re-entered the head medial superior. Similar to the results for Subject #1, the phase reversals for both SEF_{23} and SEF_{33} maps were slightly oblique from anterior superior to posterior inferior and thus orthogonal to central sulcus. Analogous to median and ulnar nerve SEFs, the magnetic field patterns of the SEF_{23} and SEF_{33} components, therefore, suggested tangential dipolar current sources in central sulcus pointing anterior for the SEF_{23} components and posterior for the SEF_{33} components. The isofield maps for the little finger were clearly medial superior compared to those for the thumb. Analogous to the findings in Subject #1, the maps for the digits in between showed a slight shift from lateral inferior to medial superior in the order index finger, middle finger, and ring finger.

Furthermore, it should be noted that these magnetic field patterns were orthogonal to the electric potential patterns observed on ECoG (Section 5.2.1.4) and on MEG (Section 5.2.2.3). Concerning the distance between the field maxima the magnetic fields were more widespread compared to the electric fields on ECoG, but more focal compared to the scalp-EEG patterns.

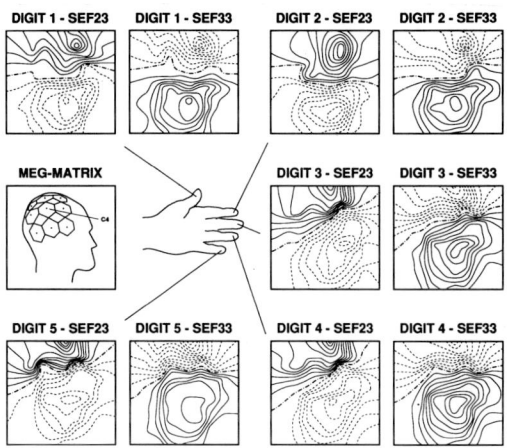

Fig. 5.2.3.3. Isofield maps for the SEF_{23} and SEF_{33} components during digit stimulation in Subject #2. Isofield maps for the SEF_{23} components reversed polarity between the upper and lower recording positions with the magnetic field emerging from the head medial superior and re-entering the head lateral inferior. Isofield maps for the SEF_{33} components showed similar patterns compared to those of the SEF_{23} components, but were inverted in orientation with the magnetic field emerging from the head lateral inferior and re-entering the head medial superior. Reversals of field direction showed a shift from lateral inferior to medial superior for digits 1 to 5. Isofield maps: 10% isocontour lines; each map scaled to maximum; solid: magnetic field emerging from the head; dashed: magnetic field re-entering the head

5.2.3.3. Cortical Hand and Digit Representation

Dipole modeling yielded good fits for a single equivalent dipole at these peak latencies explaining 83.2 ± 1.3 % of the data variance. SEF_{20} components (mixed nerve stimulation) as well as SEF_{23} components (digit stimulation) could be modeled by tangential dipoles pointing in the anterior direction; SEF_{30} components (mixed nerve stimulation) as well as SEF_{33} components (digit stimulation) could be modeled by tangential dipoles pointing in the posterior direction. The source locations during stimulation of a specific nerve or digit were close to each other at different latencies. Specifically, there was no consistent difference in the anterior-posterior and in the medial-lateral direction or concerning depth between the SEF_{20} and SEF_{23} sources on the one hand and the SEF_{30} and the SEF_{33} sources on the other.

The sources underlying median nerve stimulation were located at an average of 13.2 mm (range: 7-19 mm) lateral inferior to those underlying ulnar nerve stimulation. Similarly, there was a consistent difference between cortical thumb and little finger representation, as thumb was located at an average of 20.0 mm (range: 13-26 mm) lateral inferior to little finger. Concerning the arrangement of the digits in between, i.e. index finger, middle finger, and ring finger, in three subjects these digits followed the sensory sequence, whereas in one subject adjacent digits showed reversed cortical representations. In two subjects, thumb, index finger, and middle

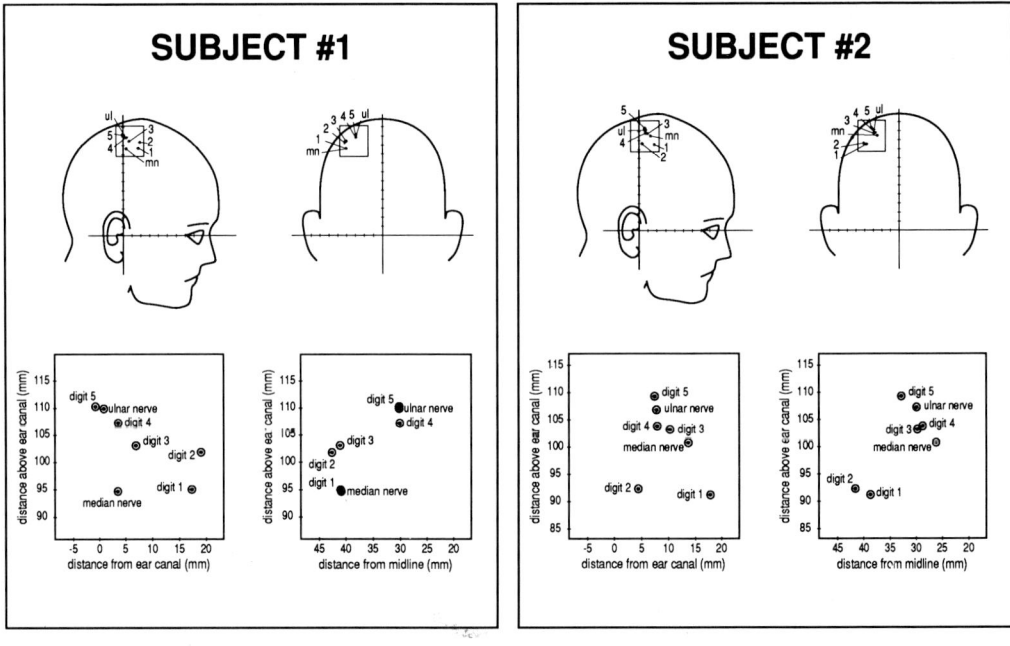

Fig. 5.2.3.4. Source locations for Subjects #1 and #2. Source locations for the SEF_{20} components (median and ulnar nerve stimulation) as well as for the SEF_{23} components (digit stimulation) are depicted as circled dots in sagittal and coronal schematic views. Regions inside boxes are enlarged at the bottom for clarity. In both subjects, the digits were arranged from lateral inferior to medial superior in the order thumb, index finger, middle finger, ring finger, and little finger. The median nerve sources were located lateral inferior to those of ulnar nerve. In Subject #1, digits 1-3 were arranged close to median nerve, whereas digits 4 and 5 were arranged close to ulnar nerve. In Subject #2, thumb and index finger were separated from the remaining cortical digit representations

finger were clustered around median nerve cortex, whereas ring finger and little finger were arranged around ulnar nerve cortex. In the other two subjects, these findings were somewhat different, as thumb and index finger were separated from middle finger, ring finger, and little finger.

In the following, we present the results of the individual subjects. The results of source localization applied to the SEF_{20} components during mixed nerve stimulation and to the SEF_{23} components during digit stimulation are shown in sagittal and coronal schematic views. Analogous to the reasoning already discussed for scalp-EEG (Section 5.2.2.4), the SEF_{20} and SEF_{23} components were chosen because at these latencies predominantly only one source is active. Thus, location estimates should be most accurate as they are not influenced by overlapping activity from other simultaneously active sources. In Subject #1 (Fig. 5.2.3.4, left), there was a clear somatotopic arrangement of cortical digit representations with a sensory sequence from lateral inferior to medial superior in the order thumb, index finger, middle finger, ring finger, and little finger. Digits 1-3 were arranged close to median nerve and were clearly separated from digits 4 and 5 which were arranged around ulnar nerve. In Subject #2 (Fig. 5.2.3.4, right), we found a similar somatotopic arrangement of the digits as in Subject #1, although there was some overlap of adjacent cortical digit representations (cf. digits 3 and 4). In contrast to Subject #1, thumb and index finger were separated from the other cortical digit representations. In Subject #3 (Fig. 5.2.3.5, left), there was some reverse of adjacent digits in the sensory sequence, i.e. middle finger was located lateral inferior to thumb and index finger, and ring finger was located medial superior to little finger. Nevertheless, there was a clear separation of median nerve and digits 1-3 which were located lateral inferior versus ulnar nerve and digits 4-5 which were located medial and superior according to the expected somatotopy. In Subject #4 (Fig. 5.2.3.5, right), finally, we could also identify a somatotopic

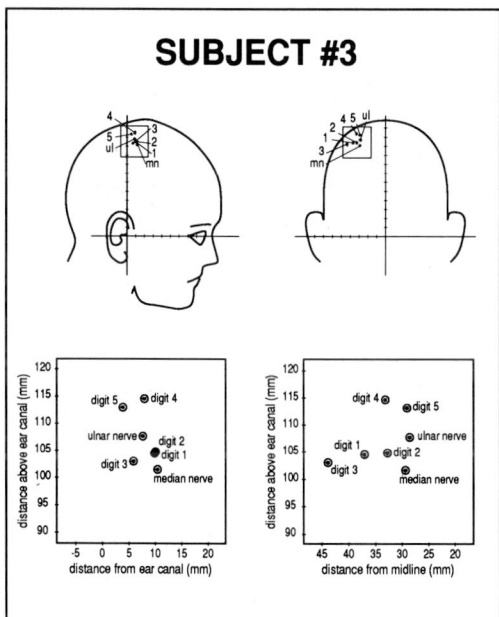
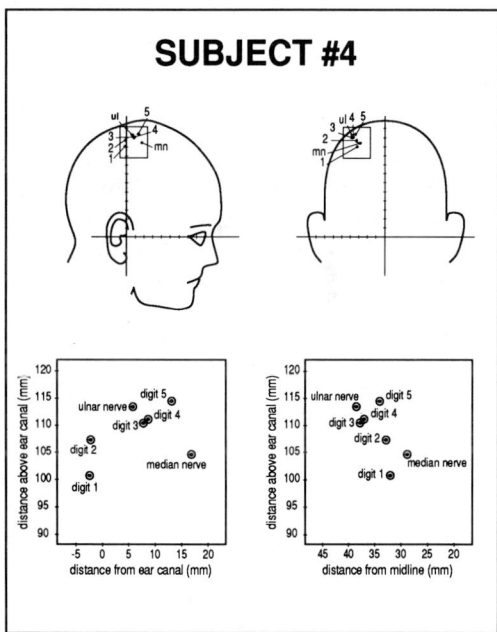

Fig. 5.2.3.5. Source locations for Subjects #3 and #4. Source locations for the SEF_{20} components (median and ulnar nerve stimulation) as well as for the SEF_{23} components (digit stimulation) are depicted as circled dots in sagittal and coronal schematic views. Regions inside boxes are enlarged at the bottom for clarity. In Subject #3, there was some reverse of adjacent digits in the sensory sequence, i.e. middle finger was located lateral inferior to thumb and index finger, and ring finger was located medial superior to little finger. Median nerve and digits 1-3 were separated from ulnar nerve and digits 4-5. In Subject #4, the digits showed a somatotopic arrangement. Median nerve and digits 1-2 were located lateral inferior and separated from ulnar nerve and digits 3-5 which were located medial superior

arrangement of cortical digit representations from lateral inferior to medial superior in the order thumb, index finger, middle finger, ring finger, and little finger. There was some overlap between middle finger and ring finger representations. Furthermore, median nerve and digits 1-2 were located lateral inferior and separated from ulnar nerve and digits 3-5 which were located medial superior.

5.2.4. Comparison of ECoG, Scalp-EEG, and MEG

5.2.4.1. General Comparison

Comparison of the raw data has been extensively discussed in Sections 4.2.4.1, 4.2.4.3, 4.2.4.4 and will not be further elucidated here. This section will focus on features specifically relevant to the study of somatotopy. It should be mentioned that we could not perform all three different recordings techniques on all subjects. In selected cases, however, we could obtain somatosensory evoked responses in the same subjects. These subjects will be presented separately in Sections 5.2.4.2 and 5.2.4.3. Here we will compare the results of the three different techniques for the subjects as a group.

Peak latencies of median, ulnar, and digit SEPs showed good agreement across techniques. Both on ECoG and scalp-EEG, P20 and N30 components (median and ulnar nerve stimulation) as well as P23 and N33 components (digit stimulation) showed tangential isopotential maps with a phase reversal between precentral and postcentral electrodes on ECoG, and between frontal and parietal electrodes on scalp-EEG. On MEG, isofield maps for the SEF_{20} and SEF_{30} components (median and ulnar nerve stimulation) as well as for the SEF_{23} and SEF_{33} components (digit stimulation) showed a phase reversal between the superior and inferior recording positions. Thus, electric and magnetic field maps were oriented orthogonal to each other. Whereas the electric phase reversals approximately followed central sulcus, the magnetic phase reversals were perpendicular to central sulcus. ECoG showed a very restricted potential pattern with an average distance of 1.7 cm between the potential maxima of the N20 and P30 as well as of the N23 and P33 components. Scalp-EEG patterns were more widespread with an average distance of 10.2 cm between the potential maxima of these components. Finally, the magnetic field patterns were more focal than the scalp-EEG patterns with an average distance of 8.9 cm between the field maxima. These findings on electric and magnetic recordings are consistent with a tangential dipolar source in the posterior wall of central sulcus.

Both on ECoG and scalp-EEG, the P25 (median and ulnar nerve stimulation) and P28 components (digit stimulation) showed radial dipolar patterns. On MEG, the occurrence of a clear SEF_{25} was inconstant across subjects. These findings, therefore, suggest a radial dipolar source in area 1 generating the 25 msec activity.

On ECoG, the separation of individual digits was clear cut and in all patients the same sensory sequence could be observed, i.e. the digits were represented from lateral inferior to medial superior in the order thumb,

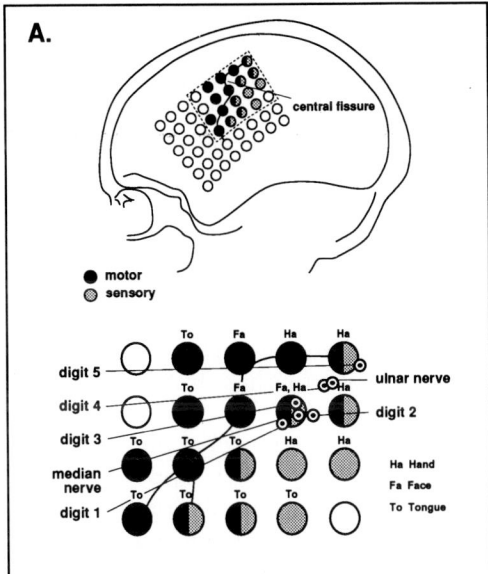
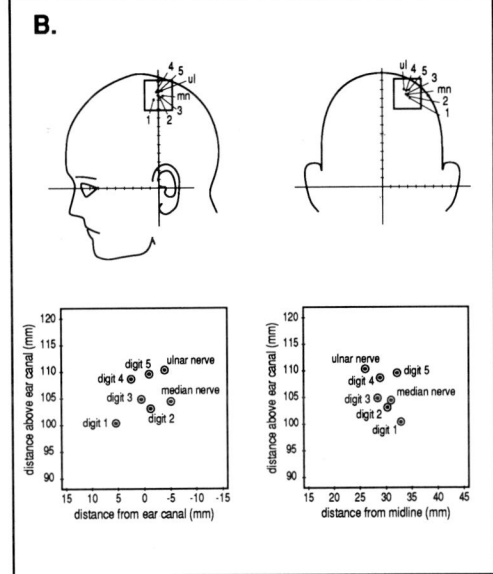

Fig. 5.2.4.1. Comparison of somatotopy on ECoG and scalp-EEG. **A.** ECoG. Skull x-ray with electrode grid (top). Region inside dotted rectangle is enlarged for clarity at the bottom. Enlargement (bottom) shows source locations and results of cortical stimulations. Motor and sensory responses elicited by cortical stimulation as well as central fissure are shown. Source locations represent surface projections and are displayed as circled dots. **B.** Scalp-EEG. Source locations are depicted as circled dots in sagittal and coronal schematic views. Boxes are enlarged at the bottom for clarity. Both on ECoG and scalp-EEG, the digits were arranged according to the sensory sequence from lateral inferior to medial superior in the order thumb, index finger, middle finger, ring finger and little finger

index finger, middle finger, ring finger and little finger. Furthermore, thumb, index finger, and middle finger were arranged around median nerve cortical representation, whereas ring finger and little finger were clustered around ulnar nerve cortex. On scalp-EEG, this sensory sequence could be replicated in two subjects, whereas in the other two subjects adjacent digits showed overlap or reversals in somatotopic arrangement. The topographical relationship of digit and mixed nerve stimulation was similar to that on ECoG in three out of 4 subjects, i.e. digits 1-3 were arranged next to median nerve and digits 4-5 were located close to ulnar nerve. In one subject, on the contrary, cortical digit representations did not show a consistent arrangement around median and ulnar nerve cortex. On MEG, the expected sensory sequence, i.e. thumb, index finger, middle finger, ring finger and little finger from lateral inferior to medial superior could be observed in three out of 4 subjects. In one subject, there was a reverse of adjacent digits. Furthermore, in two subjects digits 1-3 were separated from digits 4-5, whereas in the other two subjects digits 1-2 were separated from digits 3-5.

The distances of source locations between median and ulnar nerve were comparable for the different techniques. We found a mean distance of 9.8 mm on ECoG, of 10.8 mm on scalp-EEG, and of 13.2 mm on MEG. The distances between thumb and little finger were slightly larger and also showed good agreement across different recording methods averaging 16.1 mm on ECoG, 12.5 mm on scalp-EEG, and 20.0 mm on MEG.

5.2.4.2. Comparison of Somatotopy on ECoG and Scalp-EEG in a Selected Patient

This patient was a 30 year old female with complex partial seizures originating in the left temporal lobe. The patient's history has already been described in detail in Section 4.2.4.2. Fig. 5.2.4.1 shows the neuronal sources underlying mixed nerve and digit stimulation on ECoG as well as on scalp-EEG. On ECoG (Fig. 5.2.4.1A), source localizations are superimposed on the grid recording matrix showing the results of cortical stimulations. The sources are presented as surface projections, i.e. radial projection of the three-dimensional intracerebral source localization from the center of the sphere onto the cortical surface. On scalp-EEG (Fig. 5.2.4.1B), source localizations are depicted in a sagittal and coronal schematic view. Both on ECoG and scalp-EEG, the digits were arranged according to the sensory sequence from lateral inferior to medial superior in the order thumb, index finger, middle finger, ring finger, and little finger. The distance between the sources of thumb and little finger was 18 mm on ECoG, and 15 mm on scalp-EEG. Furthermore, both on ECoG and scalp-EEG digits 1-3 were arranged around median nerve, whereas digits 4-5 were clustered around ulnar nerve cortical representation. The distance between the median and ulnar nerve sources was 8 mm on ECoG, and 10 mm on scalp-EEG. In conclusion, scalp-EEG could approximate the results obtained from the cortical surface reasonably well.

5.2.4.3. Comparison of Somatotopy on Scalp-EEG and MEG in Two Selected Subjects

Fig. 5.2.4.2 shows the source locations obtained on Subject #A (corresponding to Subject #3 on scalp-EEG in Section 5.2.2 and to Subject #2 on MEG in Section 5.2.3). On scalp-EEG (Fig. 5.2.4.2A), digits 1-4 followed the expected sensory sequence from lateral inferior to medial superior. However, digit 5 was located close to digit 3 and thus was not represented according to the somatotopic order. Median nerve was located lateral inferior to ulnar nerve. In contrast to scalp-EEG, on MEG all digits were represented according to the sensory sequence (Fig. 5.2.4.2B). On MEG, median nerve representation was lateral inferior to that of ulnar nerve which was similar to scalp-EEG. The distance between thumb and little finger was 4 mm on scalp-EEG and 21 mm on MEG. The smaller distance on scalp-EEG resulted from the described displacement of the little finger in the sensory sequence. The distance between median and ulnar

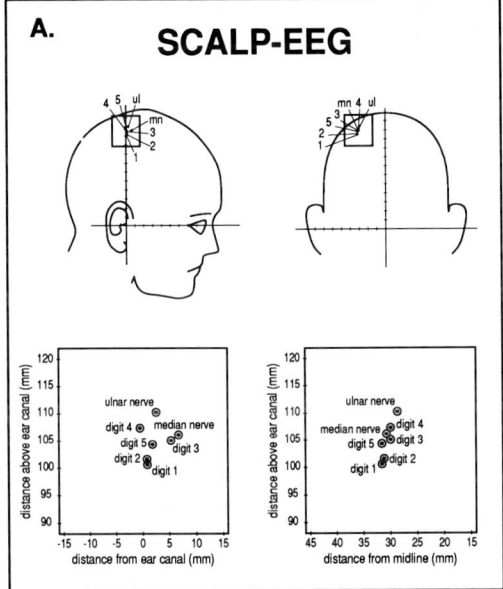
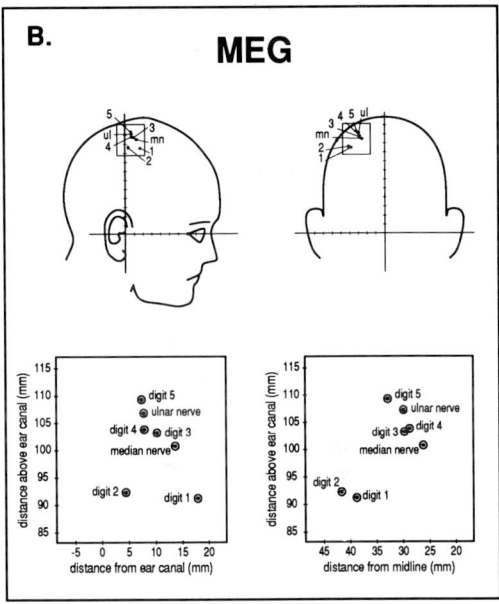

Fig. 5.2.4.2. Comparison of somatotopy on scalp-EEG and MEG for Subject #A. Source locations are depicted as circled dots in sagittal and coronal schematic views. Boxes are enlarged at the bottom for clarity. **A.** Scalp-EEG. Digits 1-4 followed the expected sensory sequence from lateral inferior to medial superior. However, digit 5 was located close to digit 3 and thus was not represented according to the somatotopic order. **B.** MEG. All digits were represented according to the sensory sequence from lateral inferior to medial superior in the order thumb, index finger, middle finger, ring finger and little finger

nerve cortex was 7 mm on scalp-EEG and 10 mm on MEG. Thus, MEG could reproduce the sensory sequence more accurate than scalp-EEG in this subject.

The results for Subject #B (corresponding to Subject #2 on scalp-EEG in Section 5.2.2 and to Subject #3 on MEG in Section 5.2.3) are shown in Fig. 5.2.4.3. In this subject, digit source localizations on scalp-EEG were arranged according to the sensory sequence (Fig. 5.2.4.3A). On MEG, adjacent digits were reversed in the sensory sequence, i.e. digit 3 was located lateral inferior to digits 1-2, and digit 4 was located lateral inferior to digit 5 (Fig. 5.2.4.3B). Both on scalp-EEG and MEG, digits 1-3 were arranged close to median nerve, and digits 4-5 close to ulnar nerve. The distance between thumb and little finger was 14 mm on scalp-EEG and 13 mm on MEG. The distance between median and ulnar nerve was 10 mm on scalp-EEG and 7 mm on MEG. In conclusion, in Subject #B the results differed from those of Subject #A as scalp-EEG could reproduce the sensory sequence more accurate than MEG.

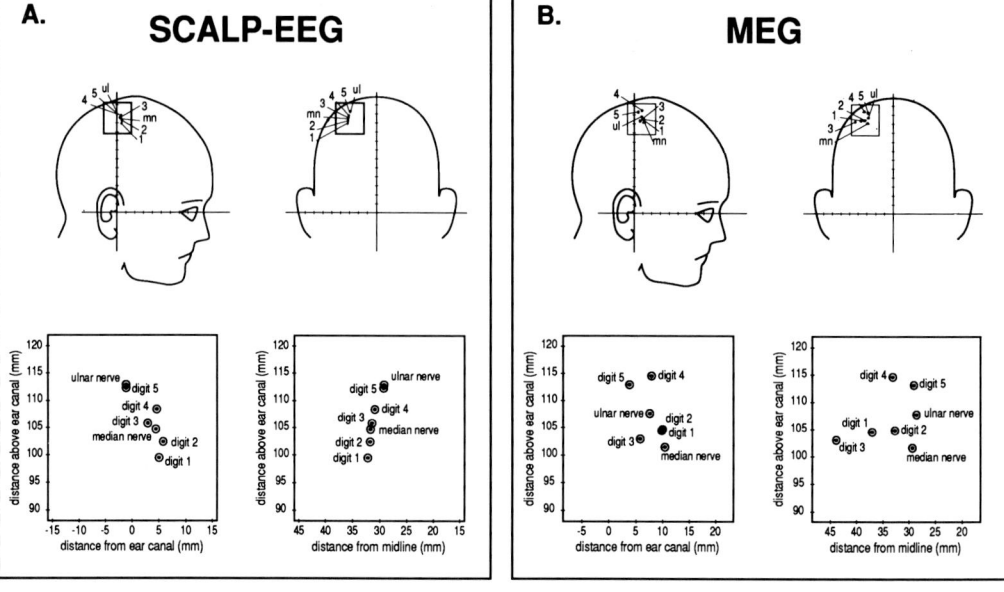

Fig. 5.2.4.3. Comparison of somatotopy on scalp-EEG and MEG for Subject #B. Source locations are depicted as circled dots in sagittal and coronal schematic views. Boxes are enlarged at the bottom for clarity. **A.** Scalp-EEG. Digit source locations were arranged according to the sensory sequence. **B.** MEG. Some adjacent digits were reversed in the sensory sequence, i.e. digit 3 was located lateral inferior to digits 1-2, and digit 4 was located lateral inferior to digit 5

5.3. Discussion

5.3.1. Somatotopy as Studied with Cortical Stimulations and Somatosensory Evoked Potentials on Electrocorticography

Our results suggest that cortical stimulations and SEPs recorded on ECoG in conjunction with dipole modeling are useful to study functional topography of human hand and lip somatosensory cortex.

5.3.1.1. Cortical Stimulations

The results of cortical stimulations have been already discussed in detail in Section 4.3.1.1. The main issues are equally applicable to this study and will be only mentioned here briefly, while the reader is referred to the Section 4.3.1.1 for a more detailed discussion. First, cortical stimulations sometimes yielded ambiguous results when stimulation of motor cortex elicited sensory responses and stimulation of somatosensory cortex elicited motor responses suggesting the concept of a sensorimotor cortex. Second, somatomotor and somatosensory cortex as outlined by cortical stimulations extended several centimeters anterior and posterior to central sulcus. Third, there was intersubject variability of the location and the spatial extent of sensorimotor cortex. Fourth, motor and sensory responses were strictly contralateral to the side of stimulation. Occasionally, ipsilateral face sensations were elicited probably due to direct stimulation of trigeminal nerve or the dura. Finally, the quality of sensation induced by cortical stimulations was usually described as tingling or numbness.

Concerning somatotopy of somatosensory cortex, our stimulation results can be summarized as follows. Stimulation of the lateral inferior electrodes elicited sensations in tongue, lip, and face. Stimulation of the more medial superior electrodes elicited sensory phenomena in the hand. Finally, stimulation of the most medial superior electrodes caused sensory responses in arm and shoulder. Responses restricted to single digits could be observed only occasionally.

These results are in agreement with the classical experiments of Penfield and Boldrey [278]. Some important aspects of this study relevant to our results are summarized in the following. As already mentioned in Section 4.3.1.1, these authors proposed a sensory sequence from lateral inferior to medial superior following central sulcus in the order: throat, jaw and teeth, taste, tongue, lips, face, nose, eye, thumb, index finger, middle finger, ring finger, small finger, hand, wrist, forearm, elbow, arm, shoulder, trunk, hip, leg (hip to foot), foot, and toes. These authors also studied the number of responses which could be achieved for each of these body parts. Whereas motor responses were most often obtained for hand, followed by lip and elbow, sensory responses could be obtained most often from hand, followed by lips, and arm.

Sensations in the hand were reported for 279 points (91 precentral), sensations in the arm (including wrist, forearm, elbow, and shoulder) for 170 points (55 precentral). The authors called attention to the remarkable spread of sensory points anterior and posterior to central fissure. Sensations in the digits could be obtained from 158 points, from 28 in front and from 130 behind of central sulcus. Compared to sensory responses for other body parts, these points were arranged in an unusual narrow strip adjacent to central sulcus. On 68 occasions responses from a single digit were obtained, on 43 from two digits, on 7 from three digits, and on 40 from all digits. Multiple finger responses were seen only in consecutive digits. Thumb responses were most frequent, followed by index finger, little finger, middle finger, and ring finger. The authors therefore concluded on a large representation of thumb compared to the other digits.

Sensations of the mouth and lip could be obtained from 74 points (21 precentral). There was a comparatively large number of precentral sensory and postcentral motor points in this group. Sensations in the tongue could be obtained from 125 points (24 precentral). Patients showed a great capacity for exact localization on the tongue and could distinguish between 4 locations, i.e. contralateral side, tip, middle, and base in most cases. The representation of the tongue extended 1.5 cm in front and behind of central sulcus. Furthermore, bilateral and ipsilateral responses were more frequently encountered than in any other part of the body. This led to a wide representation of the tongue in Penfield's charts.

Furthermore, the results of Lüders et al. [228] concerning somatotopy should be mentioned. These authors reported somatotopic features not only in the medio-lateral direction following central sulcus, but also in the anterior-posterior direction and therefore suggested the existence of somatotopically organized areas rather than a somatotopically organized strip. In agreement with these authors, we found motor and sensory responses in a band extending 3-5 cm in the rostro-caudal direction. In contrast to these authors, however, we could not observe consistent somatotopic features in the anterior-posterior direction.

5.3.1.2. Somatosensory Evoked Potentials – Data

Median and ulnar nerve SEPs showed similar features, namely tangential N20 and P30 components with phase reversals across central fissure and P25 and N35 components with single electropositivities and electronegativities, respectively. Ulnar nerve SEPs were of lower amplitude compared to median nerve SEPs. There were no significant differences concerning latencies between median and ulnar nerve SEPs. These findings here on median nerve are in agreement with previous SEP studies on ECoG [5, 7, 8, 32, 346, 378].

Digit SEPs were qualitatively similar to SEPs following mixed nerve stimulation. Thus, digit stimulation generated tangential N23 and P33 components with a phase reversal across central fissure and radial P28 and N38 components. On a quantitative basis, digit SEPs had slightly longer peak latencies and smaller peak amplitudes compared to those of median nerve. These longer peak latencies are in agreement with the results of digit SEPs on scalp-EEG [94, 95, 121, 269] and MEG [36] and can be explained by distance effects [65].

In some patients, the N20-P30 components (mixed nerve stimulation) or the N23-P33 components (digit stimulation) dominated the evoked response (cf. Patient #1). On the contrary, the P25 components (mixed nerve stimulation) or the P28 components (digit stimulation) were most prominent in other patients (cf. Patient #2). This variability is in agreement with previous ECoG studies [8, 32, 346, 378] and probably can be explained by varying contributions from different cytoarchitectonic structures [8]. However, data were reproducible within patients with low run-to-run variability. Furthermore, isopotential maps and source localizations were similar across patients which enhances reliability of our results.

Lip SEPs showed tangential N15 and P25 components with a phase reversal across central fissure which thus were analogous to median nerve N20 and P30 components. However, in contrast to hand SEPs, lip P20 and N30 – the equivalents of median P25 and N35 – showed primarily tangential patterns. SEPs during lip stimulation have been recorded on scalp EEG in normal subjects and in patients with trigeminal neuralgia [45, 55, 132, 313, 333, 334]. Different techniques have produced variable results [65]. Reports of lip SEPs on ECoG, however, have been scarce [226]. SEP waveforms and latencies of our patients were similar to those of large series on scalp-EEG [55, 333] and to those of the few reports on ECoG [226]. Previous authors, however, studied only wave morphologies and latencies with few channels, without mapping or localization of the underlying neuronal sources. We extended these studies as we mapped lip SEPs using 48 simultaneous channels, applied source localization techniques, and compared the three-dimensional locations of the neuronal sources of lip SEPs to those of median nerve, ulnar nerve, and digit SEPs.

5.3.1.3. Isopotential Maps

Isopotential maps showed similar features for median nerve, ulnar nerve, and digit SEPs, namely tangential dipolar patterns of reversed polarity for N20 and P30 (median and ulnar nerve stimulation) as well as for N23 and P33 (digit stimulation) and more radial dipolar patterns of reversed polarity for P25 and N35 (median and ulnar nerve stimulation) as well as for P28 and N38 (digit stimulation). It should be noted that N20 and P30 maps during median and ulnar nerve stimulation were – besides a reversed polarity – similar, but not identical. The same holds true for digit N23 and P33 maps. The results here for median nerve are similar to the results of other authors [8, 346, 378] and to the findings on scalp-EEG in this study (Section 5.2.2.2). The results on ulnar nerve and digits extend previous work as isopotential maps on ECoG during ulnar nerve and digit stimulation were not reported so far to our knowledge. These findings suggest a similar functional organization of cortical representation for median nerve, ulnar nerve, and the individual digits. Concerning somatotopy, median nerve isopotential maps were lateral inferior to those of ulnar nerve. Furthermore, digit isopotential maps showed a progressive shift from lateral inferior to medial superior in the order thumb, index finger, middle finger, ring finger, and little finger, which is in agreement with Penfield's sensory sequence [278, 279, 280].

Lip isopotential maps for N15 and P25 components (equivalents of median nerve N20 and P30) showed tangential dipolar patterns of opposite polarity which is in accordance with the findings during hand stimulation. However, lip N20 and P30 (equivalents of median nerve P25 and N35) showed tangential isopotential maps as well, which is different from the results of hand SEPs. This suggests differences between cortical hand and face representation which will be further elucidated in Section 5.3.1.6. Furthermore, lip isopotential maps were more widespread compared to hand isopotential maps. Concerning somatotopy, lip isopotential maps were lateral inferior to hand isopotential maps which is in accordance with the expected somatotopy of hand somatosensory cortex [278, 279, 280].

5.3.1.4. Somatotopy of Human Hand Somatosensory Cortex

Our findings confirm a somatotopic organization of cortical digit representation with a sensory sequence from sylvian fissure to vertex in the order thumb, index finger, middle finger, ring finger, and little finger. These findings of a somatotopic organization of human hand somatosensory cortex agree with previous findings in primates and humans using direct cortical stimulations and SEPs [192, 193, 194, 244, 245, 250, 278, 279, 280, 380, 381, 383]. Furthermore, we studied the cortical representations of individual digits in relation to those of median and ulnar nerve which has not been performed so far to our knowledge. Whereas thumb, index finger, and middle finger were clustered around median nerve, ring finger and little finger were arranged around ulnar nerve. Probably it is somehow artificial to model median and ulnar nerve cortical representations by dipolar point sources because stimulation of these mixed nerves should activate an extended cortical area. Thus, it is the center and not the spatial extent of activated cortex which is estimated by the dipole models applied [89, 302]. Nevertheless, our results are reasonable from a somatotopic viewpoint. Furthermore, they suggest that peripheral nerve distribution is reflected in cortical representation.

We believe that our findings are valid because the reproducibility of the results across 4 different brains is unlikely due to chance. All source localizations for individual peaks as well as the statistical means were located in postcentral gyrus within 11 mm of central sulcus which is consistent with neuronal activity in areas 3b and 1. The localization estimates agreed with the results of cortical stimulations and with anatomical features on intraoperative photographs. Averaging of a large number of trials enhanced signal-to-noise ratio.

We used a single dipole model which may have combined multiple simultaneously active sources as can be expected by multiple cortical digit representations evidenced from animal experiments [192, 193, 194, 244, 245] and from our results on spatiotemporal modeling of median nerve SEPs [32, 34, 42], see also Sections 4.2.1.4, 4.2.2.3, and 4.2.3.3. However, this is unlikely a major limitation of our results for several reasons. First, as cortical maps in areas 3b and 1 are contiguous and somatotopic in monkey [192, 193, 194, 244, 245], combining them seems reasonable to compare relative locations during stimulation of different peripheral receptive fields. Second, we were interested in the center and not the spatial extent of individual cortical digit representations. Thus, calculation of statistical means of source localizations over time seems reasonable. We think that our method – at least at the present time – is too crude to allow identification of somatotopy in different cytoarchitectonic structures. Recordings from closer space electrodes [8, 378], transcortical [8] or ideally intracortical recordings [282, 283, 291, 292], and more sophisticated physical models would be helpful to deal with these problems.

Our results extend work from previous human studies on electrocorticography using cortical stimulations and SEPs. Penfield and colleagues deduced somatotopic maps using cortical stimulation results from patient populations [278, 279, 280]. Woolsey and colleagues [380, 381, 382, 383, 384] pioneered SEP studies on electrocorticography and showed somatotopy in individual patients using ECoG amplitude criteria during finger tapping. However, these authors performed their studies in the operating room and thus could obtain measurements only from a limited number of recording sites yielding incomplete cortical maps. Furthermore, 3 patients in these studies appeared to have an occasional reverse or overlapping digit sensory sequence. We recorded SEPs from chronically indwelling subdural grid electrodes without time constraints. SEPs in our study also were not influenced by anesthesia effect on later evoked responses. We used dipole models to avoid ambiguity of peak amplitude which can be distant from a source tilted from radial due to the laws of volume conduction [83, 84, 141, 258, 337, 338, 372]. Recording from the surface of the brain is more accurate as errors due to distance effects as in MEG and scalp-EEG, and due to skull and scalp smearing effects as in scalp-EEG are avoided [1, 72, 96, 287, 346]. Comparison of our results on ECoG, scalp-EEG, and MEG, confirms this hypothesis. Digit separation was most accurate on ECoG, and MEG could separate the digits more accurately than scalp-EEG. For further details on the comparison of the different techniques see Section 5.2.4.

5.3.1.5. Neuronal Sources in Human Hand Somatosensory Cortex

The findings here on median nerve SEPs are similar to previous studies which suggested that early median nerve SEPs are generated by two neuronal sources, one in the posterior bank of the central fissure corresponding to Brodman's area 3b producing the N20-P30 components and the other in the anterior crown of the postcentral gyrus corresponding to Brodman's area 1 producing the P25-N35 components [7, 8, 32, 51, 346, 378]. Furthermore, the results here support the findings on spatiotemporal modeling in this study (Section 4.2.1).

In the present study, sources underlying median N20 and P30 components were located in postcentral gyrus and could be modeled by tangential dipolar sources located about 7.4 mm below the cortical surface. These depth estimates agree with the depth estimates of 8 mm obtained from spatiotemporal modeling in 6 patients in this study (Section 4.2.1.4). According to Allison et al. [8], the human central sulcus is approximately 18 mm deep. Thus, our results are compatible with activity at the centroid of central sulcus. The dipole locations and orientations of N20 and P30 were similar, but not identical which is in accordance with previous studies on ECoG [346] and MEG [346, 352, 375]. This can be explained by overlapping activity from the P25-N35 source as evidenced from our findings on spatiotemporal modeling (Section 4.2.1.4). Nevertheless, these results are in line with the hypothesis that the neuronal populations generating the N20 and P30 components are identical or at least similar and that these components reflect sequential depolarization of cell bodies and proximal apical dendrites (N20 component), and distal apical dendrites of cortical pyramidal cells (P30 component) [78, 212, 365, 373].

Sources underlying P25 and N35 components were located in postcentral gyrus within 11 mm of central sulcus. The depth estimates for these sources were more superficial – around 2.6 mm – compared to the depth estimates of 7.4 mm for the N20 and P30 sources. Postcentral gyrus is about 16 mm wide and area 1 occupies its anterior half [293, 361]. These findings, therefore, suggest activity in area 1 underlying the P25 and N35 components. Concerning orientations, the P25 and N35 sources were primarily radial, but also had significant tangential components. These findings agree with the results of other studies [7, 8, 32, 34, 42, 378] and with the results on spatiotemporal modeling in this study (Section 4.2.1.4). The tilt from radial can be explained in several ways. First, gyri and sulci are convoluted rather than strictly tangential or radial. Second, area 1 – besides its predominant location in surface cortex of postcentral gyrus – also occupies a small portion of the upper part of the posterior wall of central sulcus [293, 361]. Finally and probably most important, the P25 component is overlapped by activity from the N20-P30 source as suggested by Allison et al. [8] and as evidenced from our results on spatiotemporal modeling of median nerve SEPs (Section 4.2.1.4). In conclusion, it should be mentioned that the exact anatomical location of the neuronal sources generating different SEP components is still controversial. For a more detailed coverage of this question the reader is referred to Section 4.3.6.

This study extends the findings of previous studies which did not investigate the functional anatomy of ulnar nerve cortex and cortical digit representation. Like median nerve N20 and P30 components, ulnar nerve N20 and P30 components as well as digit N23 and P33 components could be modeled by tangential dipoles of opposite polarities. The depth estimates of these sources were 7.4 mm for median and ulnar nerve, and 6.1 mm for digit SEPs suggesting activity in area 3b. Furthermore, we found primarily radial dipoles for ulnar nerve P25 and N35 components as well as for digit P28 and N38 components. These sources were more superficial, i.e. at an average depth of 2.6 mm for median and ulnar nerve SEPs, and 1.0 mm for digit SEPs. These findings are compatible with activity in area 1. The reasoning for these conclusions has been already discussed in detail for median nerve cortical representation. Furthermore, these findings suggest a common cortical organization for different receptive fields of the hand.

5.3.1.6. Somatotopy and Neuronal Sources of Lip Somatosensory Cortex

Our findings on lip SEPs, however, were different from median nerve, ulnar nerve, and digit SEPs. Lip SEPs could be explained mostly by tangential sources without the requirement for additional radial sources. Similar to hand SEPs, lip N15 and P25 components – the equivalents of median nerve N20 and P30 – could be modeled by tangential dipoles of opposite polarities suggesting activity in area 3b. In contrast to hand SEPs, however, lip P20 and N30 – the equivalents of median nerve P25 and N35 – also appeared mostly explained by tangential sources which is incompatible with sole activity in area 1. The depth estimates for lip sources averaged 9.2 mm. There was no difference concerning depth between the N15-P25 sources and the P20-N30 sources which indicates that both N15-P25 and P20-N30 components reflect activity in central sulcus rather than at the cortical surface. This finding was unexpected. It suggests that lip cortical organization follows a different pattern than that of the hand. However, our results are consistent with the presence of a large cortical representation of lower lip in area 3b and a vestigial representation in area 1 of monkey [193]. The contribution of a small radially oriented cortical region could have been obscured by our use of large macroelectrodes with 6 mm diameter and 10 mm spacing center-to-center. Recordings from closer spaced electrodes [8, 378] or trans- and intracortical recordings [20, 22, 23, 111, 282, 283, 290, 291, 292] would be useful to further investigate lower lip cortical representation in area 1.

The sources in lip cortex were deeper and the dipole model accounted for less data variance than the sources in hand somatosensory cortex. Furthermore, dipole locations at different latencies showed more variability for lip than for hand SEPs. These findings are not surprising. Cortical stimulations have shown that the lips are represented in a wide area of somatosensory cortex [278, 279, 280]. SEPs during lip stimulation probably arise from an extended cortical area rather than from a focal cortical region. The widespread peaks on the lip isopotential maps are in line with this hypothesis (Figs. 5.2.1.4 and 5.2.1.5). Modeling an extended cortical area by an equivalent dipolar source results in a deeper location estimate and in a poorer dipole fit [89, 263]. Although more complex sources like dipole layers or sheets can be represented reasonably well by simple dipoles without introducing significant errors in localization on MEG and scalp-EEG [89], the dipole approximation is worse when recording in close proximity to an extended source as on ECoG in the present study. These factors probably accounted for the poorer dipole fit and deeper localizations.

5.3.2. Somatotopy as Studied on Scalp-EEG

5.3.2.1. Data

Median nerve SEPs showed P20-N30 waveforms at the frontal electrodes, and N20-P30 waveforms at the parietal electrodes with a phase reversal across central fissure. Furthermore, P25 and N35 components with single electropositivities and electronegativities could be recorded. As already discussed in more detail in Section 4.3.2.1, these findings are in agreement with previous SEP studies on scalp-EEG and ECoG [4, 5, 7, 8, 11, 51, 346, 375, 378] and the recordings from the cortical surface presented in this study (Sections 4.2.1.2). Whereas data showed good reproducibility within subjects, there was some variability across subjects similar to previous studies [4, 7, 105, 142, 346]. Low run-to-run variability as well as good reproducibility of isopotential maps and source localizations within and across subjects further support biological reliability of our results.

Ulnar nerve SEPs showed similar features compared to median nerve SEPs concerning wave-morphology, namely N20 and P30 components with phase reversals across central sulcus as well as P25 and N35 components with single electropositivities and electronegativities. Amplitudes of ulnar nerve SEPs were slightly lower compared to median nerve SEPs which can be explained by the fact that ulnar nerve contains only C8-T1 sensory afferents, whereas median nerve contains C6-T1 sensory afferents (C8-T1 muscle and C6-C8 cutaneous afferents) [65]. It has been suggested that the latencies of the Erb's point potential (EP-P) and the EP-P/N13 interpeak latency are usually longer for the ulnar nerve; presumably this is related to a longer anatomical course and entry into the spinal cord at a lower level for the ulnar nerve [65]. We could not observe significant differences in peak latencies between median and ulnar nerve SEPs which is probably due to the rather small latency differences and the small number of subjects studied.

Digit SEPs have been recorded previously by various investigators [94, 95, 100, 105, 106, 109, 110, 120, 121, 143, 153, 269]. In accordance with these studies, we found similar qualitative features for digit SEPs compared to median and ulnar nerve SEPs, i.e. N23 and P33 components with phase reversals between the frontal and parietal recording electrodes, and P28 and N38 components with single electropositivities and electronegativities, respectively. On a quantitative basis, digit SEPs showed smaller amplitudes and longer latencies compared to mixed nerve stimulation. Amplitudes of digit SEPs were about 30-50% compared to those of mixed nerve SEPs which is in accordance with other studies [95, 269]. These amplitude differences can be explained by a smaller number of fibers activated during digit stimulation [65, 269]. The observed latency differences are in agreement with the literature and can be attributed to distance effects [65, 269]. Furthermore, digit SEPs are mediated exclusively by cutaneous afferents, whereas mixed nerve stimulation is thought to recruit both muscle and cutaneous afferents [56, 134]. However, the contribution of muscle afferents to SEPs was disputed in a recent study [154]. This controversy should not affect our interpretations, however, as in the upper limbs both cutaneous and muscle afferents are believed to travel in the posterior columns [65]. This is in contrast to the lower limbs where muscle afferents travel in the faster conducting dorsolateral (spinocerebellar) pathways whereas cutaneous afferent volleys travel in the slower posterior columns [65].

Most of the aforementioned studies, however, recorded digit SEPs only from a limited, few number of electrodes and/or were not concerned with somatotopic features of hand somatosensory cortex as only selected digits were stimulated [95, 100, 105, 106, 109, 110, 143, 153]. To our knowledge, only two groups previously performed systematic multichannel scalp-EEG studies on somatotopy of digit SEPs [94, 121]. The advantages of multichannel recordings have already been mentioned in Section 1.1.2.3. These arguments hold true especially when small differences in SEP waveforms are critical for correct interpretation of the results like for differentiation of individual digits.

5.3.2.2. Isopotential Maps

Analysis of isopotential maps revealed tangential dipolar patterns of reversed polarity for the N20 and P30 components (median and ulnar nerve stimulation) as well as for the N23 and P33 components (digit stimulation) and more radial dipolar patterns of reversed polarity for P25 and N35 components (median and ulnar nerve stimulation) as well as for P28 and N38 components (digit stimulation). The findings here on median nerve and ulnar nerve are in agreement with the results of previous studies on scalp-EEG and ECoG [4, 7, 346, 375] as well as with the results on ECoG in this study (Sections 5.2.1.3, 5.2.1.4, and 5.3.1.3). Discrepancies between these median nerve SEP components and hence isopotential patterns and those of other investigators [94, 100, 109] have been extensively discussed in Sections 4.3.2.1 and 4.3.2.4.

Concerning digit SEPs, Deiber et al. [94] found tangential dipolar maps at latencies of 21-22 msec, and Onofrj et al. [269] found similar map configurations at mean latencies of 23.7 msec. These components were referred to as N20 components irrespective of their latency in these studies [94]. These findings agree with our digit N23 maps. Subsequently, these authors observed a concentric configuration of isopotential lines over central areas which was referred to as P22 component and exhibited peak latencies of 23-25 msec [94] and 25.5 msec [269], respectively. This activity was attributed to precentral structures as already discussed in detail in Sections 4.3.2.1 and 4.3.2.4. Whether the P22 components in these studies are equivalent to our P28 components cannot be determined. In any case, our interpretation of these components is different. Duff [121] presented isopotential maps only for one peak latency, i.e. for the P24 components during digit SEPs which showed radial dipolar patterns. These P24 components showed a similar configurations as our P28 maps.

In our study, we analyzed successive isopotential patterns over the entire time domain of the evoked response. Furthermore, we compared maps generated by stimulation of mixed nerve to those during digit stimulation activating primarily cutaneous afferents. Thus, we could investigate dynamical functional anatomy of hand somatosensory cortex. Our results on successive isopotential patterns and time evolution of the evoked response suggest a common cortical organization for different receptive fields of the hand, and for mixed as well as for cutaneous afferents.

5.3.2.3. Somatotopy of Human Hand Somatosensory Cortex

In contrast to previous digit SEPs studies on scalp-EEG, we applied source localization techniques to study somatotopy of the neuronal sources underlying stimulation of different peripheral receptive fields. Duff [121] reported on somatotopic features on scalp-EEG during stimulation of individual digits. He based his conclusions on topographic features of potential patterns and peak amplitude criteria, but did not apply source localization techniques. Deiber et al. [94] used a bisection method for tangential sources and maximum amplitude criteria for radial sources to infer somatotopy of different SEP components. For the N20 component, these authors could not find a somatotopic arrangement. However, dipole orientations were more horizontal for thumb and index finger and more vertical for ring finger and little finger. For the P22 component, a lateral to medial shift could be observed when changing the stimulation site from thumb to little finger. For the P27 component finally, no somatotopic differences between digits could be identified. It should be mentioned that in this study no source localization techniques based on physical models were applied. Furthermore, conclusions were based on pooled results from 12 subjects and no comments on cortical digit representation in a single brain were made.

Thus, our study is an extension of those previous reports for several reasons. First, we applied source localization techniques based on physical models. Second, we present cortical digit representation separately for each subject. Finally, we related cortical digit representations to those of median and ulnar nerve. Our results can be summarized as follows. First, median nerve cortical representation was located lateral inferior to that of ulnar nerve. The distance between median and ulnar nerve was approximately 10.8 mm which agrees with the cortical maps of Penfield [278, 279, 280] and our results on cortical stimulations and SEPs on ECoG (Section 5.2.1.6) as well as on SEFs on MEG (Section 5.2.3.3). Second, thumb, index finger, and middle finger were clustered around median nerve, whereas ring finger and little finger were arranged around ulnar nerve. This result is compatible with peripheral nerve representation. Third, digit representation seemed to follow a somatotopic pattern with a sensory sequence from lateral inferior to medial superior in the order thumb, index finger, middle finger, ring finger, and little finger [37].

However, it should be mentioned that adjacent digits occasionally showed overlap or reversal in the sensory sequence. This could be explained in several ways. First, there could be a lack of complete fractionation in the cortical representation of adjacent body parts. This hypothesis is in agreement with the studies of Woolsey and colleagues [380, 381, 383] who got similar results when recording SEPs from the cortical surface during mechanical finger tapping. The second explanation could be the lack of spatial resolution of scalp-EEG due to distance effects and smearing effects of skull and scalp [1, 72, 96, 287]. This second possibility is our working hypothesis as our ECoG recordings yielded a clear and reproducible separation for the individual digits (Section 5.2.1.6).

Some further limitations of our study should be mentioned. First, we used single equivalent dipoles to model activity probably generated by multiple simultaneously active sources as evidenced from our results of spatiotemporal modeling (Section 4.2.2.3) which may be inadequate. However, we chose the N20 components for median and ulnar nerve localization and the N23 components for digit localization representing the first cortical activity. At these latencies, spatiotemporal modeling showed that only a single cortical generator was active. Thus, the application of single dipole models at these latencies seems to be justified. At later time points, single dipole models yielded more ambiguous results, and a somatotopy of individual digits was less evident. This resulted probably from modeling multiple brain sources by a single dipole and is in agreement with the findings of Deiber et al. for their P27 components [94]. Second, we used a 4-shell spherical model. More accurate results would probably have been obtained with realistic head shape and conductivity models. However, these models are mathematically and computationally elaborate and could not be applied with the presently available hard- and software.

5.3.2.4. Neuronal Sources in Human Hand Somatosensory Cortex

Dipole modeling yielded tangential sources pointing in the anterior direction for the N20 (mixed nerve stimulation) and N23 components (digit stimulation). Furthermore, tangential sources pointing in the posterior direction were found for the P30 (mixed nerve stimulation) and P33 components (digit stimulation). Finally, the P25 (mixed nerve stimulation) and P28 components (digit stimulation) could be modeled by primarily radial sources. The dipole fits showed a good approximation of the data and explained an average of 86.1% of the variance. All sources were within 15 mm of the C3/C4 position of the International 10-20-System and thus in close proximity to primary somatosensory cortex [173, 174, 180, 331]. These findings support the hypothesis that two significant neuronal sources underlie early SEPs, namely one horizontal dipole in the posterior bank of central sulcus and one radial dipole in the anterior crown of postcentral gyrus which was initially proposed by the group around Allison, Broughton and Wood [4, 7, 8, 49, 50, 51, 375, 378] and subsequently has been confirmed by other authors [32, 34, 42, 346]. These authors, however, performed detailed studies only on median nerve SEPs. Furthermore, the findings here are in agreement with the results on ECoG and MEG on this study (Sections 4.2.1.4, 4.2.3.3, 5.2.1.6, and 5.2.3.3).

Similar to our results, Desmedt et al. [100, 105, 109] suggested neuronal sources in area 3b generating the N20 components underlying digit SEPs (equivalents of the N23 components in the present study). In contrast to our results, these authors, however, suggested a different neurogenesis for the subsequent components, specifically a radial precentral generator in motor cortex for the P22 component. Deiber et al. [94] suggested a similar neurogenesis of digit SEPs in their study on somatotopy. Subsequently, other authors followed this working hypothesis [95, 269]. Possible causes for these differences have been extensively discussed in Section 4.3.2.4.

We could not find a significant difference in depth for the tangential N20, N23, P30, and P33 sources (thought to be generated in area 3b) on the one hand, and the radial P25, P28, N35, and N38 sources (thought to be generated in area 1) on the other hand. This finding was different from the results of spatiotemporal modeling of median nerve SEPs on scalp-EEG (Section 4.2.2.3) and different from the results of digit SEPs on ECoG (Section 5.2.1.6) where the sources attributed to area 1 were located more superficially than those attributed to area 3b. This would be expected for activity in surface cortex compared to activity in the sulcus. The results here can probably be explained by inaccuracies induced by the simple spherical dipole model (see Section 2.2).

In conclusion, it should be stressed that attribution of SEP components to specific cytoarchitectonic structures certainly is beyond the accuracy limits of scalp-EEG. Nevertheless, the results here can be used to form hypotheses concerning the neurogenesis of the human somatosensory evoked response which can be tested by comparison with the results obtained on ECoG and MEG.

5.3.3. Somatotopy as Studied on Magnetoencephalography

5.3.3.1. Data

Whereas median and ulnar nerve SEFs showed prominent peaks at 20 and 30 msec (SEF_{20} and SEF_{30}), digit SEFs exhibited peak latencies at 23 and 33 msec (SEF_{23} and SEF_{33}). SEF_{20} during median and ulnar nerve stimulation reversed orientations between the upper and lower recording positions, with the magnetic field emerging from the head superior, and re-entering the head inferior. Median nerve and ulnar nerve SEF_{30} also reversed orientations between upper and lower recording positions, with the magnetic field emerging from the head inferior and re-entering the head superior. SEF_{25} could not always be clearly identified. These findings are similar to previous MEG studies on median and ulnar nerve [162, 176, 200, 346, 352, 375].

The findings on digit SEFs were qualitatively similar to those on mixed nerve stimulation, as during digit SEF_{23} the magnetic field emerged from the head superior, and re-entered the head inferior, and during digit SEF_{33} the magnetic field emerged from the head inferior and re-entered the head superior. The latency differences can be explained by distance effects [65]. The amplitude of digit SEFs was considerable smaller compared to mixed nerve stimulation which can be attributed to the smaller number of activated peripheral nerve afferents [65]. These findings of a longer latency and reduced amplitude is in agreement with the results of Kaukoranta et al. [200] who compared median nerve SEFs to those evoked by simultaneous stimulation of index and middle finger. On an absolute scale, these authors obtained longer latencies due to different filter settings. Furthermore, the results here agree with our findings on ECoG (Section 5.2.1.2) and scalp-EEG (Section 5.2.2.1) and those of other investigators on ECoG [348] and scalp-EEG [94, 95, 121, 269]. Data were reproducible both across and within subjects further enhancing biological reliability of our results.

Similar to the results on spatiotemporal modeling (Section 4.2.3.3) and to previous MEG studies [346, 352, 375], we could not always clearly identify a consistent SEF_{25} component from the raw data during median nerve stimulation (cf. Subject #1). As already mentioned in Sections 4.2.3.3 and 4.3.6.3, this activity is thought to be generated in the anterior crown of postcentral gyrus [4, 7, 8, 32, 346, 378]. MEG is sensitive primarily to tangential dipoles reflecting activity in cortical sulci, whereas it is insensitive to radial dipoles generating activity on the cortical gyri. Thus, the radial activity at 25 msec should theoretically not produce a detectable MEG signal [150, 369]. However, area 1 occupies – besides its predominant location in cortical surface cortex – also a small portion of the posterior bank of central gyrus [293, 361] resulting in a small tangential contribution to this component. It is this tangential activity that can be picked up with low noise equipment in the MEG [346, 352]. From inspection of the raw data, however, this component cannot always be seen as in Subject #1 of this study. Nevertheless, spatiotemporal modeling can help to separate and clearly identify this component in a highly reproducible way. For further details on this procedure, the reader is referred to Section 4.2.3.3. Similar findings hold true for the ulnar nerve SEF_{25} and the digit SEF_{28} components.

5.3.3.2. Isofield Maps

The following basic conclusions can be drawn from inspection of the isofield maps. First, SEF_{20} (median and ulnar nerve stimulation) and SEF_{23} (digit stimulation) as well as SEF_{30} (median and ulnar nerve stimulation) and SEF_{33} (digit stimulation) showed qualitatively similar isofield maps. Specifically, for the SEF_{20} and SEF_{23} maps the magnetic field emerged from the head superior and re-entered the head inferior with a slightly oblique phase reversal from anterior superior to posterior inferior. On the contrary, for the SEF_{30} and SEF_{33} maps the magnetic field emerged from the head inferior and re-entered the head superior with a similar phase reversal. Whereas findings here on median nerve and ulnar nerve agree with previous MEG studies [176, 346, 352, 375], the results on digit maps extend previous MEG reports on digit SEFs [48, 268, 342] as these applied the steady state paradigm and not – like in the present study – transient evoked responses. Our findings suggest similar cortical representations for different receptive fields of the hand.

Second, the phase reversal lines for median nerve were inferior lateral to those of ulnar nerve like in the study of Huttunen et al. [176]. The phase reversal lines for digit SEFs showed a progressive shift from inferior lateral to superior medial in the sequence thumb, index finger, middle finger, ring finger, and little finger which extends previous studies. These findings suggest a somatotopic arrangement of cortical digit representations.

Third, phase reversal lines were approximately perpendicular to central fissure. Furthermore, the midpoint between the field extrema was close to the C3/C4 position of the International 10-20-System which in turn is within 15 mm of central sulcus [173, 174, 180, 331]. Thus, these findings suggest tangential activity in the posterior wall of central sulcus generating these components which is in agreement with the interpretations of most previous MEG studies [34, 42, 162, 166, 176, 200, 346, 352, 375].

Finally, the maps at successive peak latencies, i.e. SEF_{20} and SEF_{30} for median and ulnar nerve as well as SEF_{23} and SEF_{33} maps for individual digits were similar, but not identical which is in accordance with previous MEG studies [200, 352, 375] and with the results here on ECoG (Sections 5.2.1.3 and 5.2.1.4) and scalp-EEG (Section 5.2.2.2 and 5.2.2.3). These differences in magnetic field patterns could result from overlapping contributions of other simultaneously active sources and will be further elucidated in the section on neuronal sources (Section 5.3.3.4).

5.3.3.3. Somatotopy of Human Hand Somatosensory Cortex

Our results show that MEG in conjunction with source localization techniques can be used to study functional somatotopy of hand somatosensory cortex non-invasively. Cortical digit representations showed a somatotopic organization with thumb represented lateral inferior and little finger medial superior. The distance between cortical thumb and little finger representation averaged 20.0 mm. Furthermore, in 3 out of 4 subjects we found a sensory sequence from sylvian fissure to vertex in the order: thumb, index finger, middle finger, ring finger, and little finger. In the fourth subject, a similar trend was found but some reverse occurred between adjacent digits. Cortical representation of median nerve was lateral inferior to ulnar nerve with an average distance of 13.2 mm. The arrangement of digits around median and ulnar nerve representations showed some differences between subjects. In two subjects, thumb, index finger, and middle finger were clearly separated from ring finger and little finger. Whereas thumb, index finger, and middle finger were clustered around median nerve, ring finger and little finger were clustered around ulnar nerve. In the other two subjects, middle finger was separated from thumb and index finger, and closer to ring finger and little finger. This may reflect interindividual differences in receptive field organization and/or cortical anatomy. Furthermore, it should be noted that the location estimates of median and ulnar nerve reflect the center of mass of an extended area of cortical activation. However, the spatial extent of activation cannot be assessed with the presently available models [368].

Our results are in agreement with previous MEG studies on digit SEFs in several aspects. Concerning the distance between cortical thumb and little finger representation, readings from Okada's [268] Table 2 suggest values of approximately 7 mm and 18 mm, respectively in the two subjects presented in detail. Brenner et al. [48] did not perform source localization techniques but observed a shift of approximately 2 cm for the isofield maps of little finger versus thumb. Finally, Suk et al. [342] did not provide data concerning the distance of thumb versus little finger. Similar to our findings in Subject #3, Okada et al. [268] and Suk et al. [342] observed reverses or overlap of thumb and index finger in some subjects. This is in agreement with the results of Woolsey et al. [380, 383] who showed somatotopy in individual patients using ECoG amplitude during tactile finger tapping. In these studies, some patients appeared to have occasional reverse or overlapping sequence of cortical digit representation. However, these findings probably result from a lack of spatial resolution of MEG, as our recordings from the cortical surface allowed complete separation of digits according to the expected sensory sequence in all subjects. This issues will be further elucidated in Section 5.3.4 on comparison of the different techniques.

Our results extend these previous MEG studies [48, 268, 342] on somatotopy of cortical digit representation for three reasons. First, we studied all individual digits, whereas these authors stimulated only selected digits: Brenner et al. [48] studied thumb and little finger, whereas Okada et al. [268] and Suk et al. [342] studied thumb, index finger, and little finger. Second, we studied the relation of cortical representation of individual digits to that of median and ulnar nerve. Finally, these studies used the steady state paradigm. On the contrary, we used transient evoked responses which allowed us to study temporal evolution and thus functional topography of human cortical hand representation.

Furthermore, our findings of a somatotopic organization of human hand somatosensory cortex agree with previous studies in primates and humans using direct cortical stimulations and SEPs [193, 278, 348, 380, 381, 383] and with the results on ECoG (Section 5.2.1.6) and scalp-EEG (5.2.2.4) in this study.

5.3.3.4. Neuronal Sources in Human Hand Somatosensory Cortex

SEF_{20} and SEF_{30} (mixed nerve stimulation) as well as SEF_{23} and SEF_{33} components could be modeled by tangential dipoles oriented in the anterior-posterior direction. Dipole modeling showed excellent fits for these components explaining an average of 83.2% of the data variance. The average depth of the sources was 22 mm below the scalp and 9 mm below the cortical surface. These depth estimates are in agreement with both SEF studies on median nerve and digits [268, 346, 352, 375]. There was no consistent difference concerning depth between the sources underlying mixed nerve stimulation and those underlying cutaneous digital nerve stimulation. These findings are consistent with activity in the posterior bank of central sulcus corresponding to Brodman's area 3b which is in accordance with previous SEF studies on MEG [34, 42, 346, 352, 375]. In this context, it is of interest to notice that motor fields (readiness magnetic fields) were localized in the anterior bank of the central sulcus corresponding to Brodman's area 4 [62, 63].

Source localization estimates for SEF_{20} and SEF_{30} as well as for SEF_{23} and SEF_{33} were close to each other, but not identical. This difference could result from overlapping activity of other, simultaneously active sources. Our results from spatiotemporal modeling suggest that this overlapping activity is generated by the SEF_{25}-SEF_{35} source in area 1 which is in agreement with the conclusions of Wood et al. [375]. However, this interpretation is different from that of the Finnish group [176, 352, 353]. These investigators found that the SEF_{30} sources were medial and anterior to the SEF_{20} sources. According to Woolsey et al. [383] the hand is represented more medially on the motor than on the sensory side of central sulcus. Thus, these investigators concluded that SEF_{30} is partly generated in motor cortex. On the contrary, we could not find a systematic difference between the SEF_{20} and SEF_{30} sources regarding the medial to lateral or the anterior to posterior direction which agrees with the results of other studies [346, 375]. From our findings, we therefore cannot support the interpretation of a contribution from motor cortex for the SEF_{30} source suggested by the Finnish group. Instead, we propose that both SEF_{20} and SEF_{30} are generated mainly in area 3b and that SEF_{30} is partly overlapped by activity from area 1.

Thus, SEF_{20} and SEF_{30} should reflect the sequential activation of proximal and distal apical dendrites of cortical pyramidal cells in area 3b on a microscopic level [8, 78].

Kaukoranta et al. [200] compared SEFs during mixed median nerve stimulation at the wrist and its cutaneous branches on the glabrous skin. The equivalent dipoles for mixed nerve stimulation were located deeper underneath the scalp than those activated by cutaneous nerve stimulation. Therefore, these authors suggested that mixed nerve stimulation activates areas 3a and 3b whereas cutaneous stimulation activates mainly area 3b. The reasoning for this hypothesis was the following. Mixed nerve stimulation activates both muscle spindle afferents and cutaneous branches. According to experiments in primates, muscle spindle afferents project to area 3a, cutaneous fibers project to areas 3b and 1, and area 2 receives afferents from joint and other deep receptors [192]. The exact anatomical location of area 3a is not clear in humans, but in monkey it is located at the bottom of central sulcus, and thus deeper than area 3b [288]. Thus, activation of areas 3a and 3b during mixed nerve stimulation would be compatible with deeper depth estimates. We could not reproduce these results as in our study there were no significant differences in depth of the sources underlying mixed nerve stimulation and digit stimulation. Furthermore, the contribution of muscle afferents to SEPs has been recently disputed by Halonen et al. [154] which makes the reasoning of these authors questionable.

5.3.4. Comparison of ECoG, Scalp-EEG, and MEG

On ECoG, scalp-EEG, and MEG, activity at 20 msec (mixed nerve stimulation) and 23 msec (digit stimulation) was generated by tangential dipolar sources pointing in the anterior direction and thus compatible with activity the posterior wall of central sulcus. Activity at 30 msec (mixed nerve stimulation) and 33 msec (digit stimulation) could be modeled by tangential dipoles of similar locations pointing in the posterior direction. The electric isopotential maps (ECoG and scalp-EEG) at these latencies were similar in configuration and were perpendicular to the magnetic isofield maps. This finding can be explained by the fact that electric and magnetic fields generated by identical neuronal processes are perpendicular to each other [69]. The activity at 25 msec (mixed nerve stimulation) and 28 msec (digit stimulation) could be modeled by radial sources both on ECoG and scalp-EEG. This activity produced only small signals on MEG as magnetic fields generated by radial dipoles cancel out in a spherical volume conductor [150]. Thus, these findings yield converging evidence for generation of the 25 msec activity in surface cortex of the postcentral gyrus. The hypothesis of two neuronal generators in the posterior bank of central sulcus and in the anterior crown of postcentral gyrus underlying early SEPs and SEFs following mixed nerve as well as digit stimulation are in accordance with the results of previous studies on median nerve [7, 8, 32, 34, 42, 345, 375, 378]. These issues are discussed in more detail in Section 4.3.6. Whereas combined neuroelectric and neuromagnetic studies have been performed on median nerve evoked responses [346, 375], digit SEPs and SEFs have not been investigated using a combined approach of these techniques. Thus, our results are an extension of these previous studies and suggest similar functional organizations for different receptive fields of the hand supported by converging evidence from neuroelectric and neuromagnetic studies.

All three techniques confirmed a somatotopic arrangement of cortical digit representation. These findings are in agreement with animal experiments and to the results of cortical stimulations in humans [192, 193, 194, 244, 245, 278, 279, 280, 380, 381, 382, 383, 384]. ECoG was most accurate as cortical digit representations were separated on ECoG in all patients, and the digits followed the sensory sequence proposed by Penfield [278, 279, 280], i.e. were arranged from lateral inferior to medial superior in the order thumb, index finger, middle finger, ring finger, and little finger. On MEG, this sensory sequence could be reproduced in three out of 4 subjects. On scalp-EEG finally, two out of 4 subjects exhibited this sensory sequence [37, 38, 39].

These findings were expected and can be explained as follows. First, ECoG should be most accurate as measurements are made directly from the cortical surface and thus ECoG does not loose spatial resolution due to distance effects. On scalp-EEG and MEG, measurements have to be performed at a distance due to the thickness of the skull and scalp resulting in a loss of accuracy [346]. Furthermore, electrical fields measured on ECoG are not distorted by the resistive properties of the skull and scalp like those measured on scalp-EEG [1, 72, 96, 287]. Second, MEG should be more accurate compared to scalp-EEG. Although MEG measurements are made at a distance similar to scalp-EEG, skull and scalp are essentially transparent to magnetic fields [69, 81, 162, 198, 199, 369, 370] whereas electric fields are attenuated and smeared [1, 72, 96, 287]. Another physical condition should favor magnetic compared to electric measurements for the study of cortical digit representation. The direction of highest sensitivity of these neurophysiological techniques is given by a line connecting the field maxima [162]. The direction of highest sensitivity on MEG follows the sensory sequence along central sulcus, as the SEF_{20} and SEF_{30} components (median and ulnar nerve stimulation) as well as the SEF_{23} and SEF_{33} components (digit stimulation) reversed in direction between the upper and lower ends of central sulcus. On the contrary, the electric fields for the N20 and P30 components (median and ulnar nerve stimulation) and for the N23 and P33 components (digit stimulation) showed potential maxima at the pre- and postcentral (ECoG) as well as at the frontal and parietal (scalp-EEG) electrodes. Thus, for electric measurements the direction of highest sensitivity is oriented perpendicular to the sensory sequence.

Despite these theoretical considerations, scalp-EEG and MEG actually complemented each other in spatial discrimination of individual digit representations and in correct identification of the sensory sequence. Whereas in some subjects MEG was superior to scalp-EEG (cf. Subject #A), in other subjects scalp-EEG was more accurate than MEG (cf. Subject #B).

This finding is in agreement with previous studies on epileptic spikes performed by Sutherling et al. [345]. In this study, some epileptiform discharges could be recorded on MEG, but not on scalp-EEG, whereas other spikes showed up on scalp-EEG, but not on MEG. Scalp-EEG yielded reasonable source locations concerning somatotopy in all subjects. Furthermore, in one patient studied with both ECoG and scalp-EEG the relative location estimates showed excellent agreement of these techniques. In conclusion, the combined use of scalp-EEG and MEG should enhance information which can be derived from non-invasive recordings and thus should improve localization accuracy.

On both scalp-EEG and MEG, we observed an occasional reverse or overlap of adjacent digits. This finding could occur due to intersubject variability in cortical anatomy or due to inaccuracies of the recording techniques. Similar findings were reported by Woolsey et al. [381, 383] during mechanically evoked responses recorded from the cortical surface. On the contrary, other authors found cortical digit representations exactly following the expected sensory sequence [348]. This is also in agreement with the results on ECoG in this study (Section 5.2.1.6). We did not perform all three recording techniques on all subjects. In one subject studied both with ECoG and scalp-EEG, the sensory sequence on scalp-EEG agreed with that obtained on ECoG. In two other subjects studied both with scalp-EEG and MEG, there were differences in cortical digit representations as obtained with these two techniques. Thus, our working hypothesis is that cortical digit representation follows an orderly somatotopic arrangement and that reverses in this sequence are due to inaccuracies of the recording technique.

Finally, the distance estimates of thumb versus little finger as well as median nerve versus ulnar nerve showed good agreement on ECoG, scalp-EEG, and MEG. Thus, scalp-EEG and MEG should be useful to measure the spatial extent of cortical hand representation non-invasively which could be of clinical relevance as discussed in Section 6.

6. Clinical Implications

Accurate localization of central fissure is important in patients undergoing surgery in regions adjacent to somatosensory and motor cortex to prevent disabling and permanent neurological deficits like severe paresis. The correct identification of primary motor cortex, somatosensory cortex, and central fissure cannot be reliably made by visual inspection of the cortical surface due to large inter-patient variability in the configuration of gyri and surface vasculature [5, 378]. Wood et al. [378] systematically reviewed cortical anatomy in 46 neurosurgical patients who were evaluated with SEPs on ECoG for localization of cortical hand area. These authors could not observe a consistent pattern concerning morphology of sulci, gyri, and overlying vasculature. While the central sulcus often made a characteristic bend in the region of the hand area, it sometimes did not, and similar bends were seen on both pre- and postcentral sulci. Furthermore, large lesions may displace common anatomical landmarks and cause tissue compression and brain edema which makes identification of sulci virtually impossible. Thus, no surface clues exist how to approach subcortical lesions which implies risk to injure the primary motor and somatosensory regions [149]. Even if there exist surface clues about the location of a subcortical lesion, it is important to know the spatial relationship of this lesion to functionally important brain areas [149]. MRI scans, intraoperative ultrasound [60], and CT guided stereotaxy [169] can be used to localize the surface area overlying a subcortical lesion and their combined use with neurophysiologic techniques should increase the safety of removing lesions from the vicinity of central sulcus [149].

Furthermore, long-standing lesions also can cause functional reorganization especially in the developing brains of children [149]. Thus, functional rather than anatomical criteria have to be used for resections in close proximity to essential cortex. This necessitates the use of neurophysiologic techniques as cortical stimulations and evoked potentials.

Cortical stimulations are considered the most accurate method for localization of essential cortex at the present time [219, 226, 228, 261, 262]. However, they are time consuming, require collaboration of the patient, and sometimes yield equivocal results [378]. Stimulation of primary motor cortex can be performed under general anesthesia. However, during general anesthesia the threshold for producing movement is raised and the extent of the electrically excitable cortex is constricted. Thus, much time is spent by varying stimulus intensity and exploring the cortical surface with the stimulation electrode [149]. Furthermore, the use of general anesthesia precludes reports of sensations [5]. Thus, adequate localization has to be performed under local anesthesia which is an unpleasant experience for the patient and the surgical team alike [5]. Furthermore, in young children motor cortex may be electrically unexcitable and SEPs thus are the only means of somatic localization [146]. The reason for this phenomenon is not clear. It does, however, not appear to be related to pathological changes in motor cortex [146]. A dissociation between the ontogenetic development of electrical excitability of motor cortex and the ability of the somatosensory cortex to generate a response to electrical stimulation may be an explanation [144]. This hypothesis is further substantiated by animal experiments [171, 175].

This has led to reliance on SEPs recorded from the cortical surface to localize motor and somatosensory cortex. Although SEPs are useful generally for localization of central fissure during neurosurgery adjacent to central fissure they are particularly valuable in some special clinical conditions according to the observations made by Wood et al [378]. In patients with medically refractory seizures arising from the lateral frontal or parietal neocortex exact delineation of motor and somatosensory cortex allows maximal resection of the epileptogenic tissue without the risk of subsequent severe motor or sensory deficits. Similarly, SEP localization allows more complete removal of tumors while sparing functional important brain tissue, especially if this tumor has lead to displacement of normal brain structures. Finally, SEPs are useful in children where neurosurgery under local anesthesia should be avoided.

We recorded SEPs on ECoG from chronically indwelling subdural grid electrodes. Nevertheless, it should be mentioned that recording of SEPs provides equally good localization under local and general anesthesia in the operating room. For instance, Allison et al. [5] used 40-60 % nitrous oxide in oxygen supplemented with 0.25-0.5 % isoflurane or a similar halogenated anesthetic and observed no significant effects on SEPs. Higher concentrations of halogenated agents, however, may depress or abolish SEPs.

Although SEPs on ECoG have been recorded by several groups, objective criteria for localization of central fissure on ECoG have been provided only by few investigators [5, 146, 149, 378]. Allison et al. [5] and Wood et al. [378] provided three different criteria for localization of central sulcus. First, the phase reversal for the N20 and P30 components across central sulcus. In this context, the authors stressed the geometric relationship of SEPs to central sulcus as follows. Due to a bend of central sulcus, the hand area of the somatosensory cortex formed a convex cap of tissue pointing towards the frontal midline. Thus, an 'on-axis' line connecting the maxima of the P20 and N20, as well as the maxima of the N30 and P30, respectively, formed an acute angle ($70° \pm 12°$) with a line approximating the overall course of central sulcus (the 'CS-line'). The authors pointed towards the importance of recording along this 'on-axis' because otherwise the frontal potential maxima could be easily missed and false localizations would be obtained. The second criterion suggested by this group was the peak amplitude of the P25 component which was usually located on the postcentral gyrus about 1 cm medial to the focus of the 20- and 30 ms potentials. The third criterion, finally, was the largest SEP amplitude regardless of specific wave morphology and peak identification which could be recorded from the hand representation area on the precentral and postcentral gyri. The authors used examination of the raw data or plots of overall amplitude to localize the maximum amplitude focus.

Goldring and Gregorie [146, 149] used bipolar recordings for localization of central sulcus. As already discussed in Section 4.3.1.2, these authors obtained responses invariably from the primary somatosensory cortex and variably from the motor cortex. Phase reversals were used as localization criteria. In the case of a sole response from somatosensory cortex, only two adjacent pairs of electrodes showed responses of reversed polarity. In the case of a response both from somatosensory and motor cortex, four adjacent pairs of electrodes showed responses and there was a double phase reversal.

Spatiotemporal modeling thus could be useful as an additional method for localization of primary motor cortex and somatosensory cortex. There are several advantages of this approach. First, inspection of peak amplitudes or phase reversals is sometimes ambiguous and subjective. Furthermore, it can sometimes be difficult to distinguish precentral P25-N35 from precentral P20-N30 components [378]. On the contrary, dipole modeling takes into account the biophysical laws of propagation of electrical fields in the brain and thus should provide more accurate and objective localization estimates. Furthermore, the source localizations can be superimposed unto the grid recording matrix. Second, the localization estimates of the N20-P30 and the P25-N35 source were separated by approximately 10 mm. Thus, this approach is useful to provide quantitative estimates of the spatial extent of cortical hand representation. Third, spatiotemporal modeling could separate the neuronal sources of SEPs in area 3b and 1. This differentiation could be of clinical relevance for the following reason. Gregorie et al. [149] reported on a patient with a tumor in right posterior frontal lobe. Resection involved a portion of the posterior bank of central sulcus in hand somatosensory area. This patient postoperatively exhibited weakness of the left upper extremity which recovered rapidly to normal function. The authors explained this finding by the fact that the resection involved primarily area 1 whereas area 3b was spared. This is in line with animal experiments which have shown that area 3b is the most important area for learning of somatosensory discrimination [298]. Thus, spatiotemporal modeling could be useful to delineate these individual cytoarchitectonic structures. However, we think that at the present time it is to early to make final conclusions on the value of spatiotemporal modeling in a clinical setting. Further studies on more patients and comparison with the established techniques are warranted to investigate its utility in difficult cases.

A special note should be made here on the clinical usefulness of recording digit and lip SEPs. Although central fissure near hand motor cortex is best identified using median nerve SEPs, the spatial extent of hand somatosensory cortex can be outlined by this technique only to a limited degree, i.e. by comparing source locations of the N20-P30 and P25-N35 sources. Thus, digit SEPs should be useful to outline cortical hand representation more exactly both in the medial-lateral and in the anterior-posterior direction. Lip SEPs may be useful for differentiating cortical hand and face representation. In the series of Gregorie et al. [149], patients with resection of the entire inferior portion of the precentral (somatomotor) gyrus had complete recovery of facial movements. This was in contrast to patients with resections involving hand motor cortex who experienced significantly worse recoveries. Furthermore, resections in hand somatosensory cortex can lead to permanent loss of fine finger movements especially if area 3b is not spared. These different recovery functions for resections involving cortical face versus hand representation could be explained by the greater bilateral cortical representation of the face than of the extremities. Differentiation of hand and lip somatosensory cortex could especially be useful in planning surgical resections for medically intractable seizures arising from operculum. Furthermore, mapping out a complete homunculus in somatosensory cortex could provide benchmarks of representations of different body parts in primary motor cortex. However, it has to be considered that the cortical representations in primary motor and somatosensory cortex are not simply mirror images from each other but show some shifts in the medio-lateral direction according to importance. Thus Woolsey et al. [383] observed that the face-arm boundary was situated more laterally on postcentral than on precentral gyrus. These authors explained this finding of an upward shift of the face-arm boundary precentrally by a greater need for cortical representation of motor than of sensory functions of the head. Nevertheless, the present results should be useful to provide additional objective criteria for functional localization of human primary motor and somatosensory cortex. A complete homunculus study is probably difficult to obtain in the operating room due to time constraints and thus is reserved to an extraoperative setting necessitating chronic grid implantation. However, a special protocol could be designed for the operating room, e.g. stimulation of median nerve, thumb, little finger, and lip.

The disadvantage of both cortical stimulations as well as of SEPs recorded from the cortical surface is their invasive character. Thus, these techniques are restricted to intraoperative settings or to special situations in the presurgical evaluation of epilepsy patients where subdural grids are implanted for definitive localization of the seizure focus. However, the invasiveness of these techniques precludes their use in routine patient evaluation. Therefore, spatiotemporal modeling on scalp-EEG and MEG may provide additional benefits in a large number of patients. First, the location of a lesion in relation to essential cortex could be established preoperatively. This could help to decide whether to operate on the lesion at all and furthermore could guide the surgical approach. Second, it could be decided preoperatively if hand somatosensory cortex is in the region of a planned craniotomy. Third, the information derived from non-invasive techniques should also be beneficial for intraoperative SEP recordings to help correct placement of the grid. According to Wood et al. [378] recording of SEPs on ECoG takes about 15 min using a 64 electrode array if the initial estimate of hand somatosensory cortex is correct. If the initial guess is incorrect, the array has to be moved and the recording procedure has to be repeated until somatosensory cortex is identified. Each repetition takes about 5 – 10 min. Thus, operation time could be shortened considerably if correct initial estimates were provided by scalp-EEG and MEG. Fourth, scalp-EEG and MEG could be used to map out a complete detailed homunculus preoperatively and thus could yield valuable resection guidelines for the neurosurgeon. Finally, the non-invasive study of functional anatomy should also be beneficial in a large population of patients with cerebrovascular disease or tumors suffering form hemiparesis. In these patients, the neuronal substrates of functional recovery and neuronal plasticity under various forms of therapy could be studied. Especially, somatotopic studies of the change or reorganization of somatosensory cortex over time would be interesting in these patients. This might have implications to guide or develop therapeutic strategies. Non-invasive studies should benefit from the combined use of scalp-EEG and MEG as these techniques yield both complementary and confirmatory information to each other.

So far, no definitive criteria for non-invasive localization of hand somatosensory cortex with scalp-EEG and MEG have been provided to our knowledge. Thus, our results should be useful to establish such criteria. A next logical step in this direction would necessitate the correlation of source locations with structural information obtained on CT or MRI scans. In the long-term future, these neurophysiologic techniques could provide a non-invasive functional topography of essential brain areas which could be related to structural lesions. This would be especially important with the advent of new neurosurgical procedures like stereotactic radiosurgery [216, 232].

Finally, neurophysiologic techniques should not be limited to identification of primary motor and somatosensory cortex. The ultimate goal should be a complete functional anatomy of the cerebral cortex, i.e. localization of language cortex and association cortex in the anesthetized patient [149]. Gregorie et al. [149] speculated that different cyto-architectonic structures might have specific electrophysiologic markers. This hypothesis is supported by findings on the direct cortical response (DCR) which consists of an electric potential recorded in immediate vicinity of a focal electrical stimulus applied to the cortical surface [20, 22, 23]. In cat and monkey, this response is distinctly different in primary sensory areas and association areas which might reflect cortex specific differences in synaptic organization [149, 213, 336]. In any case, several different techniques like structural (CT, MRI) and functional imaging (SPECT, PET) should be combined with neurophysiologic techniques to expand our knowledge of functional anatomy of the human brain.

7. Summary

In the first part of this study, we studied functional anatomy of human somatosensory cortex using cortical stimulations and somatosensory evoked responses (SERs). SERs following median nerve stimulation were recorded on ECoG (6 patients), scalp-EEG (5 subjects), and MEG (9 subjects). We applied source localization techniques to study the neuronal sources underlying SERs and thus tried to achieve a better understanding of spatiotemporal information processing in human somatosensory cortex. Finally, we tried to provide exact and objective criteria for the localization of central fissure. We used a combined approach of multivariate statistical methods, i.e. principal component analysis, and of biophysical modeling involving multiple dipoles. The dipoles were fixed in space and had time varying activities and thus allowed modeling of multiple simultaneously active brain regions overlapping both in space and time. The basic principles of the modeling procedure were outlined in a separate section where also simple simulation experiments and typical applications were demonstrated. This approach allowed us to study the spatiotemporal structure of the evoked response over its entire time domain, and to infer the number, three-dimensional intracerebral locations, and time activities of its underlying neuronal sources. We found that two sources were sufficient to explain the large majority of the evoked response on all three recording techniques accounting for 84.6 % of the data variance on ECoG, for 87.7 % on scalp-EEG, and for 86.4 % on MEG. One source had peak latencies at about 20 and 30 msec, and generated the electric N20-P30 components and the magnetic SEF_{20}-SEF_{30} components (N20-P30 and SEF_{20}-SEF_{30} source, respectively). The other source had peak latencies at about 25 and 35 msec, and thus generated the electric P25-N35 components and the magnetic SEF_{25}-SEF_{35} components (P25-N35 and SEF_{25}-SEF_{35} source, respectively). The spatiotemporal patterns of the two sources showed considerable overlap and could not be readily identified by visual inspection of the raw data. This was especially true for the magnetic recordings, where the SEF_{25}-SEF_{35} source showed only small contributions to the evoked response. Our results support the hypothesis that the N20-P30 source consists of a horizontal dipole in the posterior wall of central sulcus, and that the P25-N35 source is represented by a radial dipole in the anterior crown of postcentral gyrus. On ECoG, both sources were located in postcentral gyrus at an average distance of 6 mm from central sulcus close to electrodes which elicited sensory experiences in the hand. The distance between the two sources and the C3/C4 position of the International 10-20 System averaged 10 mm on scalp-EEG and 11 mm on MEG.

The N20-P30 and the SEF_{20}-SEF_{30} source was represented by a tangential current dipole oriented in the anterior-posterior direction resulting in an electric potential pattern with a phase reversal between the anterior and posterior electrodes, and an orthogonal magnetic field pattern with a phase reversal between the superior and inferior recording positions. The phase reversal lines followed central sulcus for the electric measurements and were perpendicular to central sulcus for the magnetic recordings. The depth estimates for this source were similar across techniques, i.e. 8 mm on ECoG, 10 mm on scalp-EEG, and 9 mm on MEG. These orientations and depth estimates suggest activity in the posterior wall of central sulcus.

The orientations and depth estimates for the P25-N35 and the SEF_{25}-SEF_{35} source as achieved with the different recording techniques were more variable. On ECoG, the orientation of the P25-N35 source was variable across patients, i.e. more radial in some patients and more tangential in others, and the depth estimates were similar to those for the N20-P30 source (8 mm) which could be explained with difficulties in modeling cortical surface activity on ECoG with simple dipoles. On scalp-EEG, the P25-N35 source consisted of a radial dipole which was more superficial than the N20-P30 source in all subjects, i.e. 4 mm (P25-N35 source) versus 10 mm (N20-P30 source). On MEG, the SEF_{25}-SEF_{35} source also was more superficial compared to the SEF_{20}-SEF_{30} source, i.e. 3 mm (SEF_{25}-SEF_{35} source) versus 10 mm (SEF_{20}-SEF_{30} source). The magnetic SEF_{25}-SEF_{35} source was oriented tangentially, and had a significantly smaller variance contribution than the SEF_{20}-SEF_{30} source as only the tangential components of this primarily radial source could be measured on MEG. These findings therefore are supportive of a neurogenesis of the P25-N35 and the SEF_{25}-SEF_{35} source in the anterior crown of postcentral gyrus.

In conclusion, this study helped to elucidate the neuronal sources underlying the human SER and to achieve objective localization estimates of central fissure.

In the second part of the study, we studied somatotopy of human somatosensory cortex. We recorded somatosensory evoked responses during stimulation of median nerve, ulnar nerve, and all the individual digits on ECoG (4 patients), scalp-EEG (4 subjects), and MEG (4 subjects). Following mixed nerve stimulation, we found on ECoG and scalp-EEG positive-negative P20-N30 wave sequences at the anterior electrodes and negative-positive N20-P30 waveforms at the posterior electrodes as well as single electropositive P25 and single electronegative N35 components. Digit evoked responses had smaller amplitudes and longer latencies compared to mixed nerve SEPs, but otherwise showed similar qualitative features, namely positive-negative P23-N33 components at the anterior electrodes, negative-positive N23-P33 waveforms at the posterior electrodes, and single electropositive P28 as well as single electronegative N38 components. On

MEG, we found prominent SEF_{20} and SEF_{30} components (mixed nerve stimulation) as well as SEF_{23} and SEF_{33} components (digit stimulation) with magnetic fields emerging from the head superior and re-entering the head inferior for the SEF_{20} and SEF_{23} components, and magnetic fields emerging from the head inferior and re-entering the head superior for the SEF_{30} and SEF_{33} components.

We applied source localization techniques to study somatotopy of the neuronal sources underlying the evoked responses. On ECoG, scalp-EEG, and MEG, the 20 and 30 msec activity (mixed nerve stimulation) as well as the 23 msec and 33 msec activity (digit stimulation) could be modeled by tangential dipolar sources oriented in the anterior-posterior direction. This resulted in an electric potential pattern with a phase reversal between the anterior and posterior electrodes approximately following central sulcus and an orthogonal magnetic field pattern with a phase reversal between the superior and inferior recording positions. The activity at 25 msec (mixed nerve stimulation) and 28 msec (digit stimulation) could be modeled by radial sources both on ECoG and scalp-EEG. This activity produced only small signals on MEG as magnetic fields generated by radial dipoles cancel out in a spherical volume conductor. Our findings suggest activity in the posterior wall of central fissure generating the N20-P30 and the SEF_{20}-SEF_{30} components (mixed nerve stimulation) as well as the N23-P33 and the SEF_{23}-SEF_{33} components (digit stimulation), and activity in the anterior crown of postcentral gyrus generating the P25-N35 (mixed nerve stimulation) as well as the P28-N38 components (digit stimulation). Furthermore, our results demonstrate similar cortical organizations for different receptive fields of the hand.

We found a somatotopic arrangement of cortical digit representations with a sensory sequence from lateral inferior to medial superior in the order thumb, index finger, middle finger, ring finger, and little finger. On ECoG, we could observe this sensory sequence in all patients and individual digits were well separated from each other. Furthermore, digit sources were located in postcentral gyrus following central fissure and were next to electrodes which produced sensory responses in the hand. On scalp-EEG, the sensory sequence could be observed in two subjects, whereas in the other two subjects adjacent digits showed overlap or reversals in the somatotopic arrangement. On MEG, three out of 4 subjects showed a somatotopic cortical digit representation whereas in one subject there was a reverse of adjacent digits. The distance between thumb and little finger averaged 16.1 mm on ECoG, 12.5 mm on scalp-EEG, and 20.0 mm on MEG.

Furthermore, we studied cortical digit representations in relation to each other and in relation to mixed nerve sources. On ECoG, we found that thumb, index finger, and middle finger were arranged around median nerve, and separated from ring finger and little finger which were located close to ulnar nerve in all patients. On scalp-EEG, we found a similar topographical relationship of digit representations to those of mixed nerves in three out of 4 subjects, whereas in one subject cortical digit representations did not show a consistent arrangement around median and ulnar nerve. On MEG, digits 1-3 were separated from digits 4 and 5 in two subjects, whereas in the other two subjects digits 1-2 were separated from digits 3-5. The distance of median and ulnar nerve sources averaged 9.8 mm on ECoG, 10.8 mm on scalp-EEG, and 13.2 mm on MEG.

Furthermore, we compared functional topography of human lip somatosensory cortex to that of hand somatosensory cortex using SEPs on ECoG in 3 epilepsy patients. We stimulated median nerve, ulnar nerve, and the lower lips. The sources were located in postcentral gyrus and showed a somatotopic organization from lateral inferior to medial superior in the order lower lip, median nerve, and ulnar nerve. The source localizations agreed with the results of cortical stimulations. The cortical regions of median and ulnar nerve each could be modeled by sequential tangential and radial dipoles. On the contrary, the central representations of lip were different and could be explained mostly by tangential dipoles. These findings suggest differences in organization of human lip and hand sensory cortex, and are consistent with a larger representation of lip in the posterior bank of central fissure in area 3b than on the gyral surface in area 1.

Clinically, our results should be useful to delineate exactly somatosensory cortex in patients undergoing resective surgery adjacent to central fissure. This could allow more complete removal of epileptogenic tissue or tumors without the danger of inducing permanent and disabling neurological deficits, and thus could help to improve surgical outcome in these patients. Furthermore, this approach could be used to study functional reorganization of the human brain and thus could help to develop new therapeutic strategies in restorative neurology.

8. List of Abbreviations

SEF	–	somatosensory evoked magnetic field
SEP	–	somatosensory evoked potential
SER	–	somatosensory evoked reponse
ECoG	–	electrocorticography
MEG	–	magnetoencephalography
PCA	–	principal component analysis

9. References

1. Abraham, K., and Ajmone-Marsan, C.: Patterns of cortical discharges and their relation to routine scalp electroencephalography. *Electroenceph. Clin. Neurophysiol.* 10: 447-461, 1958.
2. Achim, A., Richer, F., and Saint-Hilaire, J. M.: Methods for separating temporally overlapping sources of neuroelectric data. *Brain Topography* 1: 22-28, 1988.
3. Aine, C. J., George, J. S., Medvick, P. A., Supek, S., Flynn, E. R., and Bodis-Wollner, I.: Identifcation of multiple sources in transient evoked neuromagnetic responses. In: Williamson, S. J., Hoke, M., Stroink, G., and Kotani, M. (Eds.) *Advances in Biomagnetism.* Plenum, New York, 1989: 193-196.
4. Allison, T.: Scalp and cortical recordings of inital somatosensory cortex activity to median nerve stimulation in man. *Ann. NY Acad. Sci.* 388: 671-677, 1982.
5. Allison, T.: Localization of sensorimotor cortex in neurosurgery by recording of somatosensory evoked potentials. *Yale J. Biol. Med.* 60: 143-150, 1987.
6. Allison, T., Goff, W. R., and Sterman, M. B.: Cerebral somatosensory responses evoked during sleep in the cat. *Electroenceph. Clin. Neurophysiol.* 21: 461-468, 1966.
7. Allison, T., Goff, W. R., Williamson, P. D., and Van Gilder, J. C.: On the neural origin of early components of the human somatosensory evoked potential. In: Desmedt, J. E. (Ed.) *Clinical Uses of Cerebral, Brainstem and Spinal Somatosensory Evoked Potentials. Prog. Clin. Neurophysiol., Vol. 7.* Karger, Basel, 1980: 51-68.
8. Allison, T., McCarthy, G., Wood, C. C., Darcey, T. M., Spencer, D. D., and Williamson, P. D.: Human cortical potentials evoked by stimulation of the median nerve. I. Cytoarchitectonic areas generating short-latency activity. *J. Neurophysiol.* 62: 694-710, 1989.
9. Allison, T., McCarthy, G., Wood, C. C., Goff, W. R., Spencer, D. D., and Williamson, P. D.: SEPs recorded from the human cortical surface before and after resection of the somatosensory hand representation area. *Electroenceph. Clin. Neurophysiol.* 58: 45P, 1984.
10. Allison, T., Wood, C. C., and McCarthy, G.: Somatosensory evoked potentials following surgical excision of somatosesnory or motor cortex in the monkey. *Soc. Neurosci. Abstr.* 12: 1432, 1986.
11. Allison, T., Wood, C. C., Spencer, D. D., Goff, W. R., McCarthy, G., and Williamson, P. D.: Localization of sensorimotor cortex in surgery by SEP recording. *Electroenceph. Clin. Neurophysiol.* 61: S74, 1985.
12. Anziska, B., and Cracco, R. Q.: Short latency SEPs to median nerve stimulation: Comparison of recording methods and origin of components. *Electrocenceph. Clin. Neurophysiol.* 52: 531-539, 1981.
13. Arezzo, J., Legatt, A. D., and Vaughan, H. G. J.: Topography and intracranial sources of somatosensory evoked potentials in the monkey. I. Early components. *Electroenceph. Clin. Neurophysiol.* 46: 155-172, 1979.

14. Arezzo, J., Legatt, A. D., and Vaughan, H. G. J.: Topography and intracranial sources of somatosensory evoked potentials in the monkey. II. Cortical components. *Electroenceph. Clin. Neurophysiol.* 51: 1-18, 1981.

15. Ary, J. P., Klein, S. A., and Fender, D. H.: Location of sources of evoked scalp potentials: corrections for skull and scalp thickness. *IEEE Trans. Biomed. Engng.* BME-28: 447-452, 1981.

16. Asanuma, H., and Fernandez, J. J.: Organization of projection from the thalamic relay nuclei to the motor cortex in the cat. *Brain Res.* 71: 515-522, 1974.

17. Asanuma, H., Larsen, K. D., and Yumiya, H.: Receptive fields of thalamic neurons projecting to the motor cortex in the cat. *Brain Res.* 172: 217-228, 1979.

18. Asanuma, H., Larsen, K., and Yumiya, H.: Peripheral input pathways to monkey motor cortex. *Exp. Brain Res.* 38: 349-355, 1980.

19. Baleydier, C., and Mauguière, F.: Projections of the ascending somesthetic pathways to cat superior colliculus visualized by HRP. *Exp. Brain Res.* 31: 43-50, 1978.

20. Barth, D. S., Baumgartner, C., and Di, S.: Laminar interaction in rat motor cortex during cyclical excitability changes of the penicillin focus. *Brain Res.* 508: 105-117, 1989.

21. Barth, D. S., Baumgartner, C., and Sutherling, W. W.: Neuromagnetic field modeling of multiple brain regions producing interictal spikes in human epilepsy. *Electroenceph. Clin. Neurophysiol.* 73: 389-402, 1989.

22. Barth, D. S., Di, S., and Baumgartner, C.: Laminar cortical interactions during epileptic spikes studied with principal component analysis and physiological modeling. *Brain Res.* 484: 13-35, 1989.

23. Barth, D. S., and Sutherling, W. W.: Current source-density and neuromagnetic analysis of the direct cortical response in rat cortex. *Brain Res.* 450: 280-294, 1988.

24. Barth, D. S., Sutherling, W. W., and Beatty, J.: Fast and slow magnetic phenomena in focal epileptic seizures. *Science* 226: 855-857, 1984.

25. Barth, D. S., Sutherling, W. W., and Beatty, J.: Animal neuromagnetometry and its specific application to the study of focal epileptic phenomena. In: Weinberg, H., Stroink, G., and Katila, T. (Eds.) *Biomagnetism: Applications and Theory.* Pergamon Press, New York, 1985: 237-248.

26. Barth, D. S., Sutherling, W. W., and Beatty, J.: Intracellular currents of interictal penicillin spikes: Evidence from neuromagnetic mapping. *Brain Res.* 368: 36-48, 1986.

27. Barth, D. S., Sutherling, W. W., Engel, J., and Beatty, J.: Neuromagnetic localization of epileptiform spike activity in the human brain. *Science* 218: 891-894, 1982.

28. Barth, D. S., Sutherling, W. W., Engel, J., and Beatty, J.: Neuromagnetic evidence of spatially distributed sources underlying epileptiform spikes in the human brain. *Science* 223: 293-296, 1984.

29. Barth, D. S., Sutherling, W. W., Engel, J., and Beatty, J.: Neuromagnetic localization of single and multiple sources underlying epileptiform spikes in the human brain. In: Porter, R. J., Mattson, R. H., Ward, A. A., and Dam, M. (Eds.) *Advances in Epileptology: XVth Epilepsy International Symposium.* Raven Press, New York, 1984: 379-384.

30. Basar, E.: *EEG-Brain Dynamics: Relation Between EEG and Brain Evoked Potentials.* Elsevier, Amsterdam, 1980.

31. Baule, G. M., and McFee, R.: Detection of the magnetic field of the heart. *Am. Heart J.* 66: 95-96, 1963.

32. Baumgartner, C., Barth, D. S., Levesque, M. F., and Sutherling, W. W.: Functional anatomy of human hand sensorimotor cortex from spatiotemporal analysis of electrocorticography. *Electroenceph. Clin. Neurophysiol.* 78: 56-65, 1991.

33. Baumgartner, C., Barth, D. S., Levesque, M. F., and Sutherling, W. W.: Human hand and lip sensorimotor cortex as studied on electrocorticography. *Electroenceph. Clin. Neurophysiol.* 84: 115-126, 1992.

34. Baumgartner, C., Barth, D. S., and Sutherling, W. W.: Spatiotemporal modeling of somatosensory evoked magnetic fields. In: Williamson, S. J., Hoke, M., Stroink, G., and Kotani, M. (Eds.) *Advances in Biomagnetism.* Plenum Press, New York – London, 1989: 161-164.

35. Baumgartner, C., and Deecke, L.: Magnetoencephalography in clinical epileptology and epilepsy research. *Brain Topography* 2: 203-219, 1990.

36. Baumgartner, C., Doppelbauer, A., Deecke, L., Barth, D. S., Zeitlhofer, J., Lindinger, G., and Sutherling, W. W.: Neuromagnetic investigation of somatotopy of human hand somatosensory cortex. *Exp. Brain Res.* 87: 641-648, 1991.

37. Baumgartner, C., Doppelbauer, A., Sutherling, W. W., Levesque, M. F., Zeitlhofer, J., and Deecke, L.: Somatotopy of human hand somatosensory cortex as studied on scalp-EEG. *Electroenceph. Clin. Neurophysiol.* (submitted).

38. Baumgartner, C., Doppelbauer, A., Sutherling, W. W., Zeitlhofer, J., Lindinger, G., and Deecke, L.: Combined neuromagnetic and neuroelectric study of human cortical digit representation. In: Hoke, M. (Ed.) *Advances in Biomagnetism 91'.* Elsevier, Amsterdam (in press).

39. Baumgartner, C., Doppelbauer, A., Sutherling, W. W., Zeitlhofer, J., Lindinger, J., Lind, C., and Deecke, L.: Human somatosensory cortical finger representation as studied by combined neuromagnetic and neuroelectric measurements. *Neurosci. Lett.* 134: 103-108, 1991.

40. Baumgartner, C., Sutherling, W. W., Di, S., and Barth, D. S.: Investigation of multiple simultaneously active brain sources in the electroencephalogram. *J. Neurosci. Meth.* 30: 175-184, 1989.

41. Baumgartner, C., Sutherling, W. W., Di, S., and Barth, D. S.: Multiple source modeling of the human epileptic spike complex in the magnetoencephalogram. In: Williamson, S. J., Hoke, M., Stroink, G., and Kotani, M. (Eds.) *Advances in Biomagnetism.* Plenum Press, New York – London, 1989: 299-302.

42. Baumgartner, C., Sutherling, W. W., Di, S., and Barth, D. S.: Spatiotemporal modeling of cerebral evoked magnetic fields to median nerve stimulation. *Electroenceph. Clin. Neurophysiol.* 79: 27-35, 1991.

43. Baumgartner, C., Zeitlhofer, J., Lindinger, G., and Deecke, L.: Magnetoencephalographie – neurophysiologische Grundlagen und klinische Anwendungen. *EEG Labor* 13: 3-21, 1991.

44. Beatty, J., Barth, D. S., and Sutherling, W. W.: Magnetically localizing the sources of epileptic discharges within the human brain. *Naval Research Reviews* 36: 20-28, 1984.

45. Bennett, M. H., and Jannetta, P. J.: Trigeminal evoked potentials in humans. *Electroenceph. Clin. Neurophysiol.* 48: 517-526, 1980.

46. Björkeland, M., and Boivie, J.: Anatomical study of the projections from the dorsal column nuclei to the midbrain of in cat. *Anat. Embryol.* 170: 29-43, 1983.

47. Brazier, M. A. B.: *Electrical Activity of the Nervous System, 4th ed.* William and Wilkins Company, Baltimore, 1977.

48. Brenner, D., Lipton, J., Kaufman, L., and Williamson, S. J.: Somatically evoked magnetic fields of the human brain. *Science* 199: 81-83, 1978.

49. Broughton, R. J.: *Somatosensory Evoked Potentials in Man: Cortical and Scalp Recordings.* Thesis, McGill University, Montreal, Quebec, Canada, 1967.

50. Broughton, R. J.: Discussion. In: Donchin, E., and Lindsley, D. B. (Eds.) *Average Evoked Potentials. NASA SP-191.* Government Printing Office, Washington, D.C., 1969: 79-84.

51. Broughton, R., Rasmussen, T., and Branch, C.: Scalp and direct cortical recordings of somatosensory evoked potentials in man (circa 1967). *Can. J. Psychol.* 35: 136-158, 1981.

52. Bruno, A. C., and Costa Ribero, P.: Designing planar gradiometer arrays: preliminary considerations. In: Williamson, S. J., Hoke, M., Stroink, G., and Kotani, M. (Eds.) *Advances in Biomagnetism.* Plenum Press, New York – London, 1989: 649-652.

53. Buchanan, D. S., Paulson, D., and Williamson, S. J.: Instrumentation for clinical applications of neuromagnetism. In: Fast, R. W. (Ed.) *Advances on Cryogenic Engineering.* Plenum Press, New York, 1987.

54. Buchsbaum, M. S., Rigal, F., Coppola, R., Cappelletti, J., King, C., and Johnson, J.: A new system for gray-level surface distribution maps of electrical activity. *Electroenceph. Clin. Neurophysiol.* 53: 237-242, 1982.

55. Buettner, U. W., Petruch, F., Scheglmann, K., and Stöhr, M.: Diagnostic significance of cortical somatosensory evoked potentials following trigeminal nerve stimulation. In: Courjon, J., Mauguiere, F., and Revol, M. (Eds.) *Clinical Applications of Evoked Potentials in Neurology.* Raven Press, New York, 1982: 339-345.

56. Burke, D., and Skuse, N. F.: Cutaneous and muscle afferent components of the cerebral potential evoked by electrical stimulation of human peripheral nerves. *Electroenceph. Clin. Neurophysiol.* 51: 579-588, 1981.

57. Carelli, P., and Foglietti, V.: Behavior of a multiloop DC superconducting quantum inference device. *J. Appl. Phys.* 53: 7592-7598, 1982.

58. Carelli, P., and Foglietti, V.: A second-derivative gradiometer integrated with a DC superconducting interferometer. *J. Appl. Phys.* 54: 6065-6067, 1983.

59. Celesia, G.: Somatosensory evoked potentials recorded directly from human thalamus and Sm I cortical area. *Arch. Neurol.* 36: 399-405, 1979.

60. Chandler, W. F., Knake, J. E., and McGillicuddy, J. E.: Intraoperative use of real-time ultrasonography in neurosurgery. *J. Neurosurg.* 57: 157-163, 1982.

61. Cheron, G., and Borenstein, S.: Specific gating of the early somatosensory evoked potentials during active movement. *Electroenceph. Clin. Neurophysiol.* 67: 537-548, 1987.

62. Cheyne, D., Kristeva, R., and Deecke, L.: Homuncular organization of human motor cortex as indicated by neuromagnetic recordings. *Neurosci. Lett.* 122: 17-20, 1991.

63. Cheyne, D., Kristeva, R., Lang, W., Lindinger, G., and Deecke, L.: Neuromagnetic localization of sensorimotor cortex sources associated with voluntary movements in humans. In: Williamson, S. J., Hoke, M., Stroink, G., and Kotani, M. (Eds.) *Advances in Biomagnetism.* Plenum Press, New York – London, 1989: 177-180.

64. Cheyne, D., and Weinberg, H.: Neuromagnetic fields accompanying unilateral finger movements: pre-movement and movement evoked fields. *Exp. Brain Res.* 78: 604-612, 1989.

65. Chiappa, K. H.: *Evoked Potentials in Clinical Medicine.* Raven Press, New York, 1983: 203-250.

66. Chiappa, K. H., Choi, S. K., and Youn, R. R.: Short latency SEPs following median nerve stimulation in patients with neurologic lesions. In: Desmedt, J. E. (Ed.) *Clinical Uses of Cerebral, Brainstem and Spinal Somatosensory Evoked Potentials. Prog. Clin. Neurophysiol., Vol. 7.* Karger, Basel, 1980: 264-281.

67. Clark, W. E., and Powell, T. P. S.: On the thalamo-cortical connexions of the general sensory cortex of *Macaca. Proc. R. Soc. (Biol. Sci.)* 141: 467-487, 1953.

68. Cohen, D.: Magnetoencephalography: Evidence of magnetic fields produced by the alpha-rhythm currents. *Science* 161: 784-786, 1968.

69. Cohen, D., and Cuffin, B. N.: Demonstration of useful differences between the magnetoencephalogram and electroencephalogram. *Electroenceph. Clin. Neurophysiol.* 56: 38-51, 1983.

70. Cohen, D., and Hosaka, H.: Magnetic fields produced by a current dipole. *J. Electrocardiol.* 9: 409-417, 1976.

71. Cohen, L. G., and Starr, A.: Localization, timing, and specificity of gating of somatosensory evoked potentials during active movement in man. *Brain 110:* 451-467, 1987.

72. Cooper, R., Winter, A. L., Crow, H. J., and Walter, W. G.: Comparison of subcortical, cortical and scalp activity using chronically indwelling electrodes in man. *Electroenceph. Clin. Neurophysiol.* 18: 217-228, 1965.

73. Cracco, R. Q.: The initial positive potential of the human scalp recorded somatosensory evoked reponse. *Electroenceph. Clin. Neurophysiol.* 32: 623-629, 1972.

74. Cracco, R. Q.: Traveling waves of the human scalp recorded somatosensory evoked responses: effects of differences in recording technique and sleep on somatosensory and somatomotor responses. *Electroenceph. Clin. Neurophysiol.* 33: 557-566, 1972.

75. Cracco, R. Q.: Scalp-recorded potentials evoked by median nerve stimulation: subcortical potentials, traveling waves and somatomotor potentials. In: Desmedt, J. E. (Ed.) *Clinical Uses of Cerebral, Brainstem and Spinal Somatosensory Evoked Potentials. Prog. Clin. Neurophysiol., Vol. 7.* Karger, Basel, 1980: 1-14.

76. Cracco, R. Q., Anziska, B. J., Cracco, J. B., Vas, G. A., Rossini, P. M., and Maccabee, P. J.: Short-latency somatosensory evoked potentials to median and peroneal nerve stimulation: studies in normal subjects and patients with neurologic disease. *Ann. NY Acad. Sci.* 388: 412-425, 1982.

77. Cracco, R. Q., and Cracco, J. B.: Somatosensory evoked potentials in man: far-field potentials. *Electroenceph. Clin. Neurophysiol.* 41: 460-466, 1976.

78. Creutzfeldt, O., and Houchin, J.: Neuronal basis of EEG-waves. In: Remond, A. (Ed.) *Handbook of Electroencephalography and Clinical Neurophysiology, 2C.* Elsevier, Amsterdam, 1974: 5-55.

79. Cuffin, B. N.: A comparison of moving dipole inverse solutions using EEG's and MEG's. *IEEE Trans. Biomed. Engng.* BME-32: 905-910, 1985.

80. Cuffin, B. N.: Effects of fissures in the brain on electroencephalograms and magnetoencephalograms. *J. Appl. Phys.* 57: 146-153, 1985.

81. Cuffin, B. N., and Cohen, D.: Comparison of the magnetoencephalogram and electroencephalogram. *Electroenceph. Clin. Neurophysiol.* 47: 132-146, 1979.

82. Cuffin, B. N., and Cohen, D.: Effects of detector coil size and configuration on measurements of the magnetoencephalogram. *J. Appl. Phys.* 54: 3589-3594, 1983.

83. Darcey, T.: *Methods for Localization of Electrical Sources in the Human Brain and Applications to the Visual System.* Ph.D. Thesis, California Institute of Technology, 1979.

84. Darcey, T. M., Ary, J. P., and Fender, D. H.: Methods for localization of electrical sources in the human brain. In: Kornhuber, H. H., and Deecke, L. (Eds.) *Progress in Brain Research.* 1980: 128-134.

85. Dawson, G. D.: Cerebral responses to electrical stimulation of peripheral nerve in man. *J. Neurol. Neurosurg. Psychiat.* 10: 134-140, 1947.

86. Dawson, G. D.: Investigation of patients subject to myoclonic seizures after sensory stimulation. *J. Neurol. Neurosurg. Psychiat.* 10: 141-162, 1947.

87. Dawson, G. D.: Auto-correlation and automatic integration. *Electroenceph. Clin. Neurophysiol.* Suppl. 4: 26-37, 1954.

88. de Munck, J. C.: *A Mathematical and Physical Interpretation of the Electromagnetic Field of the Brain.* PhD, University of Amsterdam, 1989.

89. de Munck, J. C., Van Dijk, B. W., and Spekreijse, H.: Mathematical dipoles are adequate to describe realistic generators of human brain activity. *IEEE Trans. Biomed. Engng.* BME-25: 421-429, 1988.

90. Deecke, L.: Bereitschaftspotential as an indicator of movement preparation in supplementary motor area and motor cortex. In: Porter, R. (Ed.) *Motor Areas of the Cerebral Cortex.* Wiley, Chichester, 1987: 231-250.

91. Deecke, L., Boschert, J., Brickett, P., and Weinberg, H.: Magnetoencephalographic evidence for possible supplementary motor participation in human voluntary movement. In: Weinberg, H., Stroink, G., and Katila, T. (Eds.) *Biomagnetism: Applications and Theory.* Pergamon Press, New York, 1985: 369-372.

92. Deecke, L., Boschert, J., Weinberg, H., and Brickett, P.: Magnetic fields of the human brain (Bereitschaftsmagnetfeld) preceding voluntary foot and toe movements. *Exp. Brain Res.* 52: 81-86, 1983.

93. Deecke, L., Weinberg, H., and Brickett, P.: Magnetic fields of the human brain accompanying voluntary movement: Bereitschaftsmagnetfeld. *Exp. Brain Res.* 48: 144-148, 1982.

94. Deiber, M. P., Giard, M. H., and Mauguiere, F.: Separate generators with distinct orientations for N20 and P22 somatosensory evoked potentials to finger stimulation? *Electroenceph. Clin. Neurophysiol.* 65: 321-334, 1986.

95. Delberghe, X., Mavroudakis, N., Zegers de Beyl, D., and Brunko, E.: The effect of stimulus frequency om post- and pre-central short-latency somatosensory evoked potentials (SEPs). *Electroenceph. Clin. Neurophysiol.* 77: 86-92, 1990.

96. DeLucchi, M. R., Garoutte, B., and Aird, R. B.: The scalp as an electroencephalographic averager. *Electroenceph. Clin. Neurophysiol.* 14: 191-196, 1962.

97. Desmedt, J. E.: Non-invasive analysis of the spinal cord generators activated by somatosensory input in man: nearfield and farfield components. *Exp. Brain Res.* 19 (Suppl): 45-62, 1984.

98. Desmedt, J. E.: Generator sources of SEP in man. In: Cracco, R. Q., and Bodis-Wolner, I. (Eds.) *Evoked Potentials.* Alan R. Liss, Inc., New York, 1986: 235-245.

99. Desmedt, J. E.: Exteroceptive input to motor cortex in man. In: Struppler, A., and Weindl, A. (Eds.) *Clinical Aspects of Sensory Motor Integration.* Springer, Berlin, Heidelberg, 1987: 49-57.

100. Desmedt, J. E., and Bourguet, M.: Color imaging of parietal and frontal somatosensory potential fields evoked by stimulation of median and posterior tibial nerve in man. *Electroenceph. Clin. Neurophysiol.* 62: 1-17, 1985.

101. Desmedt, J. E., and Brunko, E.: Functional organization of far-field and cortical components of somatosensory evoked potentials in normal adults. In: Desmedt, J. E. (Ed.) *Clinical Uses of Cerebral, Brainstem and Spinal Somatosensory Evoked Potentials. Prog. Clin. Neurophysiol., Vol. 7.* Karger, Basel, 1980: 27-50.

102. Desmedt, J. E., Brunko, E., Debecker, J., and Carmeliet, J.: The system bandpass required to avoid distortion of early components when averaging somatosensory evoked potentials. *Electroenceph. Clin. Neurophysiol.* 37: 407-410, 1974.

103. Desmedt, J. E., Chalklin, V., and Tomberg, C.: Emulation of somatosensory evoked potential (SEP) components with the 3-shell head model and the problem of 'ghost potential fields' when using an average reference in brain mapping. *Electroenceph. Clin. Neurophysiol.* 77: 243-258, 1990.

104. Desmedt, J. E., and Cheron, E.: Central somatosensory conduction in man: neural generators and interpeak latencies of the far-field components recorded from neck and right or left scalp earlobes. *Electroenceph. Clin. Neurophysiol.* 50: 382-403, 1980.

105. Desmedt, J. E., and Cheron, E.: Non-cephalic reference recording of early SEPs to finger stimulation in adult or aging normal man: differentiation of widespread N18 and contralateral N20 from the prerolandic P22 and N30 components. *Electroenceph. Clin. Neurophysiol.* 553-570, 1981.

106. Desmedt, J. E., and Cheron, G.: Somatosensory evoked potentials to finger stimulation in healthy octogenarians and in young adults: wave forms, scalp topography and transit times of parietal and frontal components. *Electroenceph. Clin. Neurophysiol.* 50: 404-425, 1980.

107. Desmedt, J. E., and Cheron, G.: Somatosensory evoked potentials in man: subcortical and cortical components and their neural basis. *Ann. NY Acad. Sci.* 388: 388-411, 1982.

108. Desmedt, J. E., and Nguyen, T. H.: Bit-mapped color imaging of the potential fields of propagated and segmental subcortical components of SEPs in man. *Electroenceph. Clin. Neurophysiol.* 58: 481-497, 1984.

109. Desmedt, J. E., Nguyen, T. H., and Bourguet, M.: Bit-mapped color imaging of human evoked potentials with reference to the N20, P22, P27 and N30 somatosensory responses. *Electroenceph. Clin. Neurophysiol.* 68: 1-19, 1987.

110. Desmedt, J. E., and Tomberg, C.: Mapping early somatosensory evoked potentials in selective attention: critical evaluation of control conditions used for titrating by difference the cognitive P30, P40, P100 and N140. *Electroenceph. Clin. Neurophysiol.* 74: 321-346, 1989.

111. Di, S., Baumgartner, C., and Barth, D. S.: Laminar analysis of extracellular field potentials in rat vibrissa/barrel cortex. *J. Neurophysiol.* 63: 832-840, 1990.

112. Dillon, W. R., and Goldstein, M.: *Multivariate Analysis.* Wiley, New York, 1984.

113. Dinner, D. S., Lueders, H., Lesser, R. P., Morris, H. H., and Hahn, J.: Definition of rolandic fissure by cortical stimulations and somatosensory evoked potentials. *15th Epilepsy International Symposium,* September 28. 1983.

114. Domino, E. F., Matsuoka, S., Waltz, J., and Cooper, I.: Simultaneous recordings of scalp and epidural somatosensory evoked responses in man. *Science* 145: 1199-1200, 1964.

115. Domino, E. F., Matsuoka, S., Waltz, J., and Cooper, I.: Effect of cryogenic thalamic lesions on the somesthetic response in man. *Electroenceph. Clin. Neurophysiol.* 19: 127-138, 1965.

116. Donchin, E.: A multivariate approach to the analysis of average evoked potentials. *IEEE Trans. Biomed. Eng.* BME-13: 131-139, 1966.

117. Donchin, E., Callaway, E., Cooper, R., Demedt, J. E., Goff, W. R., Hillyard, S. A., and Sutton, S.: Publication criteria for studies of evoked potentials in man. In: Desmedt, J. E. (Ed.) *Attention, Voluntary Contraction and Event-Related Cerebral Potentials. Prog. Clin. Neurophysiol., Vol. 1.* Karger, Basel, 1977: 1-11.

118. Draper, N., and Smith, H.: *Applied Regression Analysis.* Wiley, New York, 1981.

119. Dreifuss, F. E.: Goals of surgery for epilepsy. In: Engel, J. Jr. (Ed.) *Surgical Treatment of the Epilepsies.* Raven Press, New York, 1987: 31-50.

120. Duff, T. A.: Multichannel topographic analysis of human somatosensory evoked reponses. In: Desmedt, J. E. (Ed.) *Clinical Uses of Cerebral, Brainstem and Spinal Somatosensory Evoked Potentials. Prog. Clin. Neurophysiol., Vol. 7.* Karger, Basel, 1980: 69-86.

121. Duff, T. A.: Topography of scalp recorded potentials evoked by stimulation of the digits. *Electroenceph. Clin. Neurophysiol.* 49: 452-460, 1980.

122. Duffy, F., Burchfiel, J. L., and Lombroso, C. T.: Brain electrical activity mapping (BEAM): a method for extending the clinical utility of EEG and evoked potentials data. *Ann. Neurol.* 5: 309-321, 1978.

123. Duret, D., and Karp, P.: Figure of merit and spatial resolution of superconducting flux transformers. *J. Appl. Phys.* 56: 1762-1768, 1984.

124. Dykes, R. W.: The anatomy and physiology of the somatic sensory cortical regions. *Prog. Neurobiol.* 10: 33-88, 1978.

125. Eisen, A., and Aminoff, M. J.: Somatosensory evoked potentials. In: Aminoff, M. J. (Ed.) *Electrodiagnosis in Clinical Neurology.* Churchill Livingston, New York, 1986: 535-574.

126. Elger, C. E., and Speckmann, E. J.: Focal interictal epileptiform discharges (FIED) in the epicortical EEG and their relations to spinal field potentials in the rat. *Electroenceph. Clin. Neurophysiol.* 48: 447-460, 1980.

127. Elger, C. E., and Speckmann, E. J.: Penicillin-induced epileptic foci in the motor cortex: Vertical inhibition. *Electroenceph. Clin. Neurophysiol.* 56: 604-622, 1983.

128. Elger, C. E., and Speckmann, E. J.: Mechanisms controlling the spatial extent of epileptic foci. In: Wieser, H. G., Speckmann, E. J., and Engel. J. Jr. (Eds.) *Current Problems in Epilepsy: The Epileptic Focus.* John Libbey, London, 1987: 45-58.

129. Elger, C. E., Speckmann, E. J., Prohaska, O., and Caspers, H.: Pattern of intracortical potential distribution during focal interictal epileptiform discharges (FIED) and its relation to spinal field potentials in the rat. *Electroenceph. Clin. Neurophysiol.* 51: 393-402, 1981.

130. Emerson, R. G., and Pedley, T. A.: Generator sources of median somatosensory evoked potentials. *J. Clin. Neurophysiol.* 1: 203-218, 1984.

131. Engel, J. Jr.: Approaches to localization of the epileptogenic lesion. In: Engel, J. Jr. (Ed.) *Surgical Treatment of the Epilepsies.* Raven Press, New York, 1987: 75-96.

132. Findler, G., and Feinsod, M.: Sensory evoked response to electrical stimulation of the trigeminal nerve in humans. *J. Neurosurg.* 56: 545-549, 1982.

133. Freeman, W. J.: *Mass Action in the Nervous System.* Academic Press, New York-San Francisco, 1975.

134. Gandevia, S. C., Burke, D., and McKeon, B.: The projection of muscle afferents from the hand to cerebral cortex in man. *Brain* 107: 1-13, 1984.

135. Garcia Larrea, L., and Mauguière, F.: Latency and amplitude abnormalities of the scalp farfield P14 to median nerve stimulation in multiple sclerosis. *Electroenceph. Clin. Neurophysiol.* 71: 180-186, 1988.

136. Geddes, L. A., and Baker, L. E.: The specific resistance of biological materials – a compendium of data for the biomedical engineer and physiologist. *Med. Biol. Engng.* 5: 271-293, 1967.

137. Geselowitz, D. B.: On the magnetic field generated outside an inhomogenous volume conductor by internal current sources. *IEEE Trans. Mag.* MAG-6: 346-347, 1970.

138. Gevins, A. S.: Analysis of the electromagnetic signals of the human brain: Milestones, obstacles, and goals. *IEEE Trans. Biomed. Engng.* BME-31: 833-850, 1984.

139. Giblin, D. R.: Somatosensory evoked potentials in healthy subjects and in patients with lesions of the nervous system. *Ann. NY Acad. Sci.* 112: 93-112, 1964.

140. Glaser, E. M., and Ruchkin, D. S.: *Principles of Neurobiological Signal Analysis.* Academic Press, New York, 1976.

141. Gloor, P.: Neuronal generators and the problem of localization in electroencephalography: application of volume conductor theory to electroencephalography. *J. Clin. Neurophysiol.* 2: 327-354, 1985.

142. Goff, G. D., Matsumiya, Y., Allison, T., and Goff, W. R.: The scalp topography of human somatosensory and auditory evoked potentials. *Electroenceph. Clin. Neurophysiol.* 42: 57-76, 1977.

143. Goff, W. R., Rosner, B. S., and Allison, T.: Distribution of cerebral somatosensory evoked responses in normal man. *Electroenceph. Clin. Neurophysiol.* 14: 697-713, 1962.

144. Goldring, S.: A method of surgical management of focal epilepsy, especially as it relates to children. *J. Neurosurg.* 49: 344-356, 1978.

145. Goldring, S., Aras, E., and Weber, P. C.: Comparative study of sensory input to motor cortex in animals and man. *Electroenceph. Clin. Neurophysiol.* 29: 537-550, 1970.

146. Goldring, S., and Gregorie, E. M.: Surgical management of epilepsy using epidural recordings to localize the seizure focus. *J. Neurosurg.* 60: 457-466, 1984.

147. Goldring, S., and O'Leary, J. L.: Correlation between steady transcortical potential and evoked response: I. Alterations in somatic receiving area induced by veratrine, strychnine, KCL and novocaine. *Electroenceph. Clin. Neurophysiol.* 6: 189-200, 1954.

148. Goldring, S., and Ratcheson, R.: Human motor cortex: Sensory input data from single neuron recordings. *Science* 175: 1493-1495, 1972.

149. Gregorie, E. M., and Goldring, S.: Localization of function in the excision of lesions from sensorimotor region. *J. Neurosurg.* 61: 1047-1054, 1984.

150. Grynszpan, F., and Geselowitz, D. B.: Model studies of the magnetocardiogram. *Biophys. J.* 13: 911-925, 1973.

151. Guy, C. N., Cayless, A., Walker, S., and Leedham-Green, K.: A multi-channel magnetometer. In: Williamson, S. J., Hoke, M., Stroink, G., and Kotani, M. (Eds.) *Advances in Biomagnetism.* Plenum Press, New York – London, 1989: 661-664.

152. Hahn, J. F., and Lüders, H.: Placement of subdural grid electrodes at the Cleveland Clinic. In: Engel, J. Jr. (Ed.) *Surgical Treatment of the Epilepsies.* Raven Press, New York, 1987: 621-628.

153. Halliday, A. M., and Wakefield, G. S.: Cerebral evoked potentials in patients with dissociated sensory loss. *J. Neurol. Neurosurg. Psychiat.* 26: 211-219, 1963.

154. Halonen, J. P., Jones, S., and Shawkat, F.: Contribution of cutaneous and muscle afferent fibers to cortical SEPs following median and radial nerve stimulation in man. *Electroenceph. Clin. Neurophysiol.* 71: 331-335, 1988.

155. Hämäläinen, H., Kekoni, J., Sams, M., Reinikainen, K., and Näätänen, R.: Human somatosensory evoked potentials to mechanical pulses and vibration: contributions of SI and SII somatosensory cortices to P50 and P100 component. *Electroenceph. Clin. Neurophysiol.* 75: 13-21, 1990.

156. Hämäläinen, M. S.: A 24-channel planar gradiometer: system design and analysis of neuromagnetic data. In: Williamson, S. J., Hoke, M., Stroink, G., and Kotani, M. (Eds.) *Advances in Biomagnetism.* Plenum Press, New York – London, 1989: 639-644.

157. Hämäläinen, M. S., and Sarvas, J. I.: Realistic conductivity geometry model of the human head for interpretation of neuromagnetic data. *IEEE Trans. Biomed. Engng.* BME-36:165-171, 1989.

158. Hand, P. J., and Van Winkle, T.: The efferent connections of the feline nucleus cuneatus. *J. Comp. Neurol.* 171: 83-110, 1976.

159. Haneishi, H., Ohyama, N., and Sekihara, K.: Discussion of biomagnetic imaging system and reconstruction algorithm. In: Williamson, S. J., Hoke, M., Stroink, G., and Kotani, M. (Eds.) *Advances in Biomagnetism.* Plenum Press, New York, 1989: 575-578.

160. Hari, R.: Neuromagnetic studies of the human auditory cortex: recent results. In: Atsumi, K., Kotani, M., Ueno, S., Katila, T., and Williamson, S. (Eds.) *Biomagnetism '87.* Tokyo Denki University Press, Tokyo, 1988: 34-41.

161. Hari, R., Hämäläinen, M., Ilmoniemi, R., Kaukoranta, E., Reinikainen, K., Salminen, J., Alho, K., Näätänen, R., and Sams, M.: Responses of the primary auditory cortex to pitch changes in a sequence of tone pips: neuromagnetic recordings in man. *Neurosci. Lett.* 50: 127-132, 1984.

162. Hari, R., and Kaukoranta, E.: Neuromagnetic studies of somatosensory system: principles and examples. *Prog. in Neurobiol.* 24: 233-256, 1985.

163. Hari, R., and Lounasmaa, O. V.: Recording and interpretation of cerebral magnetic fields. *Science* 244: 432-436, 1989.

164. Hari, R., Pelizzone, M., Mäkelä, J., Hällström, J., Huttunen, J., and Knuutila, J.: Neuromagnetic responses from a deaf subject to stimuli presented through a multichannel cochlear prosthesis. *Ear Hear.* 9: 148-152,1988.

165. Hari, R., Pelizzoni, M., Mäkelä, J. P., Hällström, J., Leinonen, L., and Lounasmaa, O. V.: Neuromagnetic responses of the human auditory cortex to on- and off-sets of noise bursts. *Audiology* 26: 31-43, 1987.

166. Hari, R., Renikainen, K., Kaukoranta, E., Hämäläinen, M., Ilmoniemi, R., Penttinen, A., Salminen, J., and Teszner, D.: Somatosensory evoked cerebral magnetic fields from SI and SII in man. *Electroenceph. Clin. Neurophysiol.* 57: 254-263, 1984.

167. Harman, H. H.: *Modern Factor Analysis.* University of Chicago Press, Chicago, 1967.

168. Hashimoto, S., Segawa, Y., Kawamura, J., Harada, Y., Yamamoto, T., Suenaga, T., Shigematu, K., Iwami, O., and Nakamura, M.: Volume conduction of the parietal N20 potential to the prerolandic frontal area. *Brain* 113: 1501-1509, 1990.

169. Heilbrun, M. P., Roberts, T. S., and Apuzzo, M. L. J.: Preliminary experience with Brown-Roberts-Wells (BRW) computerized tomography stereotaxic guidance system. *J. Neurosurg.* 59: 217-222, 1983.

170. Helmholtz, H.: Ueber einige Gesetze der Vertheilung elektrischer in körperlichen Leitern, mit Anwendung auf die thierelektrischen Versuche. *Ann. Phys. Chem.* 89: 211-233, 353-377, 1853.

171. Hines, M., and Boynton, E. P.: The maturation of 'excitability' in the precentral gyrus of the young monkey *(Macaca mulatta). Contrib. Embryol.* 178: 313-451, 1940.

172. Hirsch, J. F., Pertuiset, B., Calvet, J., Buisson-Ferey, J., Fischgold, H., and Scherrer, J.: Etude des reponses electrocorticales obtenues chez l'homme par des stimulations somesthesiques et visuelles. *Electroenceph. Clin. Neurophysiol.* 13: 411-424, 1961.

173. Homan, R. W.: The 10-20 electrode system and cerebral location. *Am. J. EEG Technol.* 28: 269-279, 1988.

174. Homan, R. W., Herman, J., and Purdy, P.: Cerebral location of the International 10-20 system electrode placement. *Electroenceph. Clin. Neurophysiol.* 66: 376-382, 1987.

175. Hunt, W. E., and Goldring, S.: Maturation of evoked response of the visual cortex in the postnatal rabbit. *Electroenceph. Clin. Neurophysiol.* 3: 465-471, 1951.

176. Huttunen, J., Hari, R., and Leinonen, L.: Cerebral magnetic responses to stimulation of ulnar and median nerves. *Electroenceph. Clin. Neurophysiol.* 66: 391-400, 1987.

177. Huttunen, J., Kaukoranta, E., and Hari, R.: Cerebral magnetic responses to stimulation of tibial and sural nerves. *J. Neurol. Sci.* 79: 43-54, 1987.

178. Ilmoniemi, R., Hari, R., and Reinikainen, K.: A four-channnel SQUID magnetometer for brain research. *Electroenceph. Clin. Neurophysiol.* 58: 467-473, 1984.

179. Jasper, H. H.: Electrocorticography. In: Penfield, W., and Jasper, H. H. (Eds.) *Epilepsy and the Functional Anatomy of the Brain.* Little Brown, Boston, 1954: 692-738.

180. Jasper, H. H.: The ten-twenty electrode system of the International Federation. *Electroenceph. Clin. Neurophysiol.* 10: 367-380, 1958.

181. Jasper, H., Lende, R., and Rasmussen, T.: Evoked potentials from the exposed somato-sensory cortex in man. *J. Nerv. Ment. Dis.* 130: 526-537, 1960.

182. Jones, E. G.: The nature of the afferent pathways conveying short-latency inputs to primate motor cortex. In: Desmedt, J. E. (Ed.) *Motor Control Mechanisms in Health and Disease.* Raven Press, New York, 1983: 263-285.

183. Jones, E. G.: Identification and classification of intrinsic circuit elements in the neocortex. In: Edelman, G. M., Gall, W. E., and Cowan, W. M. (Eds.) *Dynamic Aspects of Neocortical Function.* John Wiley & Sons, New York, 1984: 6-40.

184. Jones, E. G., Coulter, J. D., and Hendry, S. H. C.: Intracortical connectivity of architectonic fields in the somatic sensory, motor and parietal cortex of monkeys. *J. Comp. Neurol.* 181: 291-348, 1978.

185. Jones, E. G., Friedman, D. P., and Hendry, S. H. C.: Thalamic basis of place- and modality specific columns in monkey somatosensory cortex: a correlative anatomical and physiological study. *J. Neurophysiol.* 48: 545-568, 1982.

186. Jones, E. G., and Powell, T. P. S.: Connexions of the somatic sensory cortex of the rhesus monkey. I.-Ipsilateral cortical connexions. *Brain 92:* 477-502, 1969.

187. Jones, E. G., and Powell, T. P. S.: The cortical projection of the ventroposterior nucleus of the thalamus in the cat. *Brain Res.* 13: 298-318, 1969.

188. Jones, E. G., and Powell, T. P. S.: Connexions of the somatic sensory cortex of the Rhesus monkey. III.-Thalamic connections. *Brain* 93: 37-56, 1970.

189. Josephson, B. D.: Possible new effects in superconductive tunneling. *Phys. Letters* 1: 251-253, 1962.

190. Juliano, S. L., Friedman, D. P., and Eslin, D.: Patterns of cortico-cortical connectivity can predict patches of stimulus-evoked metabolic activity in monkey somatosensory cortex. *Soc. Neurosci. Abstr.* 13: 470, 1987.

191. Juliano, S. L., and Whistel, B. L.: Metabolic labeling associated with index finger stimulation in monkey SI: between animal variability. *Brain Res.* 342: 242-251, 1985.

192. Kaas, J. H.: The organization of somatosensory cortex in primates and other mammals. In: von Euler, C., Franzen, O., Lindblom, U., and Ottoson, D. (Eds.) *Somatosensory Mechanisms.* Macmillan Press, London, 1984: 51-59.

193. Kaas, J. H., Nelson, R. J., Sur, M., Lin, C. S., and Merzenich, M. M.: Multiple representations of the body within primary somatosensory cortex of primates. *Science* 204: 521-523, 1979.

194. Kaas, J. H., Sur, M., Nelson, R. J., and Merzenich, M. M.: The postcentral somatosensory cortex – multiple representations of the body in primates. In: Woolsey, C. N. (Ed.) *Cortical Sensory Organization.* Humana Press, Clifton, 1981: 29-45.

195. Kajola, M., Ahlfors, S., Ehnholm, G. J., Hällström, J., Hämäläinen, M. S., Ilmoniemi, R. J., Kiviranta, M., Knuutila, J., Lounasmaa, O. V., Tesche, C., and Vilkman, V.: A 24-channel magnetometer for brain research. In: Williamson, S. J., Hoke, M., Stroink, G., and Kotani, M. (Eds.) *Advances in Biomagnetism.* Plenum Press, New York – London, 1989: 673-676.

196. Katila, T. E.: Principles and applications of SQUID sensors. In: Williamson, S. J., Hoke, M., Stroink, G., and Kotani, M. (Eds.) *Advances in Biomagnetism.* Plenum Press, New York-London, 1989: 19-32.

197. Kaufman, L., Okada, Y., Brenner, D., and Williamson, S. J.: On the relation between somatic evoked potentials and fields. *Intern. J. Neuroscience* 15: 223-239, 1981.

198. Kaufman, L., Okada, Y., Tripp, J., and Weinberg, H.: Evoked neuromagnetic fields. *Ann. NY Acad. Sci.* 338: 722-742, 1980.

199. Kaufman, L., and Williamson, S. J.: The evoked magnetic field of the human brain. *Ann. NY Acad. of Sci.* 340: 45-65, 1980.

200. Kaukoranta, E., Hämäläinen, M., Sarvas, J., and Hari, R.: Mixed and sensory

nerve stimulations activate different cytoarchitectonic areas in the human primary somatosensory cortex SI. *Exp. Brain Res.* 63: 60-66, 1986.

201. Kavanagh, R. N., Darcey, T. M., and Fender, D. H.: The dimensionality of the human visual evoked scalp potenital. *Electroenceph. Clin. Neurophysiol.* 40: 633-644, 1976.

202. Kavanagh, R. N., Darcey, T. M., Lehmann, D., and Fender, D. H.: Evaluation of methods for three-dimensional localization of electrical sources in the human brain. *IEEE Trans. Biomed. Engng.* BME-25: 421-429, 1978.

203. Kelhä, V., Pukki, J. M., Peltonen, R. S., Penttinen, A. J., Ilmoniemi, R. J., and Heino, J. J.: Design, construction, and performance of a large-volume magnetic shield. *IEEE Trans. Biomed. Engng.* MAG-18: 260-270, 1982.

204. Kelly, D. L., Goldring, S., and O'Leary, J. L.: Averaged evoked somatosensory responses from exposed cortex in man. *Arch. Neurol.* 13: 1-9, 1965.

205. Ketchen, M. B.: Design of improved integrated thin-film planar DC SQUID gradiometers. *J. Appl. Phys.* 58: 4322-4325, 1985.

206. King, R. B., and Schell, G. R.: Cortical localization and monitoring during cerebral operations. *J. Neurosurg.* 67: 210-219, 1987.

207. Kirkpatrick, S., Gellatt, C. D. J., and Vecci, M. P.: Optimization by simulated annealing. *Science* 220: 670-680, 1983.

208. Klee, M., and Rall, W.: Computed potentials of cortically arranged populations of neurons. *J. Neurophysiol.* 40: 647-666, 1977.

209. Knuutila, J., Ahlfors, S., Ahonen, A., Hällström, J., Kajola, M., Lounasmaa, O. V., Vilkman, V., and Tesche, C.: Large-area low-noise seven-channel dc SQUID magnetometer for brain research. *Rev. Sci. Instrument.* 58: 2145-2156, 1987.

210. Krauskopf, J., Klemic, G., Lounasmaa, O. V., Travis, D., Kaufman, L., and Williamson, S. J.: Neuromagnetic mesaurements of visual responses to chromaticity and luminance. In: Williamson, S. J., Hoke, M., Stroink, G., and Kotani, M. (Eds.) *Advances in Biomagnetism.* Plenum, New York, 1989: 209-212.

211. Kwan, H. C., MacKay, W. A., and Murphy, J. T.: Spatial organization of the precentral cortex in awake primates. II. Motor outputs. *J. Neurophysiol.* 41: 1120-1131, 1978.

212. Landau, W. M.: Evoked potentials. In: Quarton, G. C., Melnechuk, T., and Schmitt, F. O. (Eds.) *The Neurosciences.* Rockefeller Univ. Press, New York, 1967: 469-482.

213. Landau, W. M., and Clare, M. H.: A note on the characteristic response pattern in primary sensory projection cortex of the cat following a synchronous afferent volley. *Electroenceph. Clin. Neurophysiol.* 8: 457-464, 1956.

214. Lehmann, D., and Skrandies, W.: Reference-free identification of components of checkboard-evoked multichannel potential fields. *Electroenceph. Clin. Neurophysiol.* 48: 609-621, 1980.

215. Lehnertz, K., Lütkenhöner, B., Hoke, M., and Pantev, C.: Considerations on a spatio-temporal two-dipole model. In: Williamson, S. J., Hoke, M., Stroink, G., and Kotani, M. (Eds.) *Advances in Biomagnetism.* Plenum Press, New York, 1989: 563-566.

216. Leksell, D. G.: Stereotactic radiosurgery. *Neurol. Res.* 9: 60-68, 1987.

217. Lemon, R. N., and van der Burg, J.: Short-latency peripheral inputs to thalamic neurons projecting to the motor cortex in the monkey. *Exp. Brain Res.* 36: 445-462, 1979.

218. Lesser, R. P., Lüders, H., Klem, G., Dinner, D. S., Morris, H. H., Hahn, J. F., and Wyllie, E.: Extraoperative cortical functional localization in patients with epilepsy . *J. Clin. Neurophysiol.* 4: 27-53, 1987.

219. Lesser, R. P., Lüders, H., Morris, H. H., Dinner, D. S., Klem, G., Hahn, J., and

Harrison, M.: Electrical stimulation of Wernicke's area interferes with comprehension. *Neurology* 36: 658-663, 1986.

220. Libet, B., Alberts, W. W., Wright, E. W., Dilattre, L., Levin, G., and Feinstein, B.: Production of treshold levels of conscious sensation by electrical stimulation of human somatosensory cortex. *J. Neurophysiol.* 27: 546-578, 1964.

221. Lilly, J. C.: Injury and excitation by electric currents. A. The balanced pulse-pair waveform. In: Sheer, D. E. (Eds.) *Electrical Stimulation of the Brain.* University of Austin Press, Austin, 1961: 60-64.

222. Lim, C. L., Rennie, C., Yiannikas, C., Gordon, E., Sloggett, G. J., Grognard, R. J., and Seagar, A. D.: Short latency median nerve somatosensory evoked magnetic fields and electric potentials. In: Williamson, S. J., Hoke, M., Stroink, G., and Kotani, M. (Eds.) *Advances in Biomagnetism.* Plenum Press, New York – London, 1989: 153-156.

223. Lopes da Silva, F., and Van Rotterdam, A.: Biophysical aspects of EEG and MEG generation. In: Niedermeyer, E., and Lopes da Silva, F. (Eds.) *Electroencephalography. Basic Principles, Clinical Applications and Related Fields.* Urban & Schwarzenberg, Baltimore, Munich, 1982: 15-26.

224. Lorente de Nó, R.: Action potentials of the motoneurons of the hypoglossus nucleus. *J. Cell. Comp. Physiol.* 29: 207-287, 1947.

225. Lorente de Nó, R.: *A Study of Nerve Physiology (Part 2).* Rockefeller Institute, 1947.

226. Lüders, H., Dinner, D. S., Lesser, R. P., and Morris, H. H.: Evoked potentials in cortical localization. *J. Clin. Neurophysiol.* 3: 75-84, 1986.

227. Lüders, H., Hahn, J. F., Lesser, R. P., Dinner, D. S., Rothner, D., and Erenberg, G.: Localization of epileptogenic spike foci: comparative study of closely spaced scalp electrodes, nasopharyngeal, sphenoidal, subdural, and depth electrodes. In: Akimoto, H., Kagamatsuri, H., Seino, M., and Ward, A. A. (Eds.) *Advances in Epileptology: XIII Epilepsy International Symposium.* Raven Press, New York, 1982: 185-189.

228. Lüders, H., Lesser, R. P., Dinner, D. S., Morris, H. H., Hahn, J. F., Friedman, L., Skipper, G., Wyllie, E., and Friedman, D.: Chronic intracranial recording and stimulation with subdural electrodes. In: Engel, J. Jr. (Ed.) *Surgical Treatment of the Epilepsies.* Raven Press, New York, 1987: 297-322.

229. Lüders, H., Lesser, R. P., Hahn, J., Dinner, D. S., Morris, H., Resor, S., and Harrison, M.: Basal temporal language area demonstrated by electrical stimulation. *Neurology* 36: 505-510, 1986.

230. Lüders, H., Lesser, R. P., Hahn, J., Little, J., and Klem, G.: Subcortical SEPs to median nerve stimulation. *Brain* 106: 341-372, 1983.

231. Lueders, H., Lesser, R. P., Hahn, J., Dinner, D. S., and Klem, G.: Cortical somatosensory evoked potentials in response to hand stimulation. *J. Neurosurg.* 58: 885-894, 1983.

232. Lunsford, L. D., Flickinger, J., Lindner, G., and Maitz, A.: Stereotactic radiosurgery of the brain using the first United States cobalt-60 source gamma knife. *Neurosurg.* 24: 151-159, 1989.

233. Maccabee, P. J., Pinkhasov, E. I., and Cracco, R. Q.: Short latency somatosensory evoked potentials to median nerve stimulation: effect of low frequency filter. *Electroenceph. Clin. Neurophysiol.* 55: 33-44, 1983.

234. MacIntyre, W. J., Bidder, T. G., and Rowland, V.: *The Prodcuction of Brain Lesions with Electric Currents.* Proc. Natl. Biophys. Conf. 763-832, 1959.

235. Mager, A.: Magnetisch abgeschirmte Kabine zur Aufnahme kleinster magnetischer und elektrischer Biosignale. *Naturwiss.* 69: 383-388, 1982.

236. Maier, J., Dagnelie, G., Spekreijse, H., and Van Dijk, B. W.: Principal components analysis for source localizations of VEPs in man. *Vision Res.* 27: 165-177, 1987.

237. Mauguière, F., and Desmedt, J. E.: Thalamic pain syndrome of Dejerine-Roussy: differentiation of 4 subtypes assisted by SEP data. *Arch. Neurol.* 45: 1312-1320, 1988.

238. Mauguière, F., and Desmedt, J. E.: Bilateral somatosensory evoked potentials in four patients with long-standing surgical hemispherectomy. *Ann. Neurol.* 26: 724-731, 1989.

239. Mauguière, F., Desmedt, J. E., and Courjon, J.: Astereognosis and dissociated loss of frontal and parietal components of somatosensory evoked potentials in hemispheric lesions. *Brain* 106: 271-311, 1983.

240. Mauguière, F., Desmedt, J. E., and Courjon, J.: Neural generators of N18 and P14 farfield SEPs: patients with lesion of thalamus or thalamocortical radiations. *Electroenceph. Clin. Neurophysiol.* 56: 283-292, 1983.

241. McLain, D. H.: Drawing contours from arbitrary data points. *The Computer Journal* 17: 318-324, 1974.

242. Meijs, J. W. H., Peters, M. J., and Van Osterom, A.: Computation of MEG's and EEG's using a realistically shaped multi-compartment model of the head. *Med. Biol. Eng. Comput.* 23: 36-37, 1985.

243. Meijs, J. W. H., ten Voorde, B. J., Peters, M. J., Stok, C. J., and Lopes da Silva, F. H.: The influence of various head models on EEGs and MEGs. In: Pfurtscheller, G., and Lopes da Silva, F. H. (Eds.) *Functional Brain Mapping.* Hans Huber, Toronto, 1988.

244. Merzenich, M. M., Kaas, J. H., Sur, M., and Lin, C. S.: Double representations of the body surface within cytoarchitectonic areas 3b and 1 in "SI" in the owl monkey (Aotus trivirgatus). *J. Comp. Neurol.* 181: 41-74, 1978.

245. Merzenich, M., Sur, M., Nelson, R. J., and Kaas, J. H.: Organization of the S I cortex – multiple cutaneous representations in areas 3b and 1 of the owl monkey. In: Woolsey, C. N. (Ed.) *Cortical Sensory Organization.* Humana Press, Clifton, 1981.

246. Mitzdorf, U.: Current source-density method and application in cat cerebral cortex: Investigation of evoked potentials and EEG phenomena. *Physiol. Rev.* 65: 37-100, 1985.

247. Möcks, J., and Verleger, R.: Principal component analysis of event related potentials: a note on misallocation of variance. *Electroenceph. Clin. Neurophysiol.* 65: 393-398, 1986.

248. Morrell, F., Hoeppner, T. J., and Whisler, W. W.: The use of intraopertive somatosensory evoked potentials to delineate the post-central gyrus in man. *Electroenceph. Clin. Neurophysiol.* 51: 41P, 1981.

249. Morrison, D. F.: *Multivariate Statistical Methods.* McGraw Hill, Tokyo, 1976.

250. Mountcastle, V. B.: Modality and topographic properties of single neurons of cat's somatic sensory cortex. *J. Neurophysiol.* 20: 408-434, 1957.

251. Mountcastle, V. B., Davis, P. W., and Berman, A. L.: Response properties of single neurons of cat's somatic sensory cortex to peripheral stimuli. *J. Neurophysiol.* 20: 374-407, 1957.

252. Mountcastle, V. B., and Powell, T. P. S.: Central neural mechanisms subserving position sense and kinesthesis. *Bull. Johns Hopkins Hosp.* 105: 173-200, 1959.

253. Mountcastle, V. B., and Powell, T. P. S.: Neural mechanisms subserving cutaneous sensibility, with special reference to the role of afferent inhibition in sensory perception and discrimination. *Bull. Johns Hopkins Hosp.* 105: 201-232, 1959.

254. Narici, L., Romani, G. L., Salustri, C., Pizzella, V., Torrioli, G., and Modena, I.: Neuromagnetic characterization of the cortical response to median nerve stimulation in the steady state paradigm . *Intern. J. Neurosci.* 32: 837-843, 1987.

255. Nelder, J. A., and Mead, R.: A simplex method for function minimization. *Comp. J.* 308-313, 1965.

256. Nelson, R. J., Sur, M., Felleman, D. J., and Kaas, J. H.: Representations of the body surface in postcentral parietal cortex of *Macaca fascicularis. J. Comp. Neurol.* 192: 611-643, 1980.

257. Nogueira, M. C., Brunko, E., De Rood, M., Trempont, V., and Zegers de Beyl, D.: Effects of isoflurane on pre- and postcentral short-latency somatosensory evoked potentials. *Neurology* 39: 1210-1215, 1989.

258. Nunez, P. L.: *Electric Fields of the Brain: The Neurophysics of EEG.* Oxford University Press, New York, 1981.

259. Nunez, P. L.: The brain's magnetic field: some effects of multiple sources on localization methods. *Electroenceph. Clin. Neurophysiol.* 63: 75-82, 1986.

260. Nuwer, M. R.: *Evoked Potential Monitoring in the Operating Room.* Raven Press, New York, 1986.

261. Ojemann, G. A.: Intraoperative functional mapping at the University of Washington, Seattle. In: Engel, J. Jr. (Ed.) *The Surgical Treatment of the Epilepsies.* Raven Press, New York, 1987: 635-640.

262. Ojemann, G. A., and Engel, J. Jr.: Acute and chronic intracranial recording and stimulation. In: Engel, J. Jr. (Ed.) *Surgical Treatment of the Epilepsies.* Raven Press, New York, 1987: 263-288.

263. Okada, Y. C.: Discrimination of localized and distributed current dipole sources and single and multiple sources. In: Weinberg, H., Stroink, G., and Katila, T. (Eds.) *Biomagnetism: Applications and Theory.* Pergamon Press, New York, 1985: 266-272.

264. Okada, Y. C., Kaufman, L., Brenner, D., and Williamson, S. J.: Application of a SQUID to measurements of somatically evoked fields. Transient responses to electrical stimulation of the median nerve. In: Erne, S. N., Halbohm, H. H., and Lübbig, H. (Eds.) *Biomagnetism.* Walter de Gruyter, Berlin, 1981.

265. Okada, Y. C., Lauritzen, M., and Nicholson, C.: Magnetic field associated with neural activities in an isolated cerebellum. *Brain Res.* 412: 151-155, 1987.

266. Okada, Y. C., Lauritzen, M., and Nicholson, C.: MEG source models and physiology. *Phys. Med. Biol.* 32: 43-51, 1987.

267. Okada, Y. C., and Nicholson, C.: Magnetic evoked field associated with transcortical currents in turtle cerebellum. *Biophysical J.* 53: 723-31, 1988.

268. Okada, Y. C., Tanenbaum, R., Williamson, S. J., and Kaufman, L.: Somatotopic organization of the human somatosensory cortex as revealed by neuromagnetic measurements. *Exp. Brain Res.* 56: 197-205, 1984.

269. Onofrj, M., Basciani, M., Fulgente, T., Bazzano, S., Malatesta, G., and Curatola, L.: Maps of somatosensory evoked potentials (SEPs) to mechanical (tapping) stimuli: comparison with P14, N20, P22, N30 of electrically elicited SEPs. *Electroenceph. Clin. Neurophysiol.* 77: 314-319, 1990.

270. Pantev, C., Hoke, M., Lehnertz, K., and Lütkenhöner, B.: Neuromagnetic evidence of an amplitopic organization of the human auditory cortex. *Electroenceph. Clin. Neurophysiol.* 72: 225-231, 1989.

271. Pantev, C., Hoke, M., Lehnertz, K., Lütkenhöner, B., Anogianakis, G., and Wittowski, W.: Tonotopic organization of the human auditory cortex as revealed by transient auditory evoked magnetic fields. *Electroenceph. Clin. Neurophysiol.* 69: 160-170, 1988.

272. Pantev, C., Hoke, M., Lehnertz, K., Lütkenhöner, B., Fahrendorf, G., and Stöber, U.: Identification of sources of brain neuronal activity with high spatiotemporal resolution through combination of neuromagnetic source localization (NMSL) and magnetic resonance imaging (MRI). *Electroenceph. Clin. Neurophysiol.* 75: 173-184, 1990.

273. Papakostopoulos, D., Cooper, R., and Crow, H. J.: Cortical potentials evoked by finger displacement in man. *Nature* 252: 582-584, 1974.

274. Papakostopoulos, D., Cooper, R., and Crow, H. J.: Inhibition of cortical evoked potentials and sensation by self-initiated movement in man. *Nature* 258: 321-324, 1975.

275. Papakostopoulos, D., and Crow, H. J.: Direct recording of the somatosensory evoked potentials from the cerebral cortex of man and the difference between precentral and postcentral potentials. In: Desmedt, J. E. (Ed.) *Clinical Uses of Cerebral, Brainstem and Spinal Somatosensory Evoked Potentials.* Karger, Basel, 1980: 15-26.

276. Papakostopoulos, D., and Crow, H. J.: The precentral somatosensory evoked potential. *Ann. NY Acad. Sci.* 425: 256-261, 1984.

277. Paul, R. L., Merzenich, M., and, G.: Representation of slowly and rapidly adapting cutaneous mechanoreceptors of the hand in Brodmann's area 3 and 1 of Macaca Mulatta. *Brain Res.* 36: 229-249, 1972.

278. Penfield, W., and Boldrey, E.: Somatic and sensory representation in the cerebral cortex of man as studied by electrical stimulation. *Brain* 60: 389-443, 1937.

279. Penfield, W., and Jasper, J.: *Epilepsy and the Functional Anatomy of the Brain.* Little Brown & Co, Boston, 1954.

280. Penfield, W., and Rasmussen, T.: *Cerebral Cortex of Man: A Clinical Study of Localization of Function.* Macmillan Co, New York, 1950.

281. Petsche, H.: Die Architektonik der Großhirnrinde (Neokortex). In: Zenker, W. (Ed.) *Benninghoff: Makroskopische und mikroskopische Anatomie des Menschen, 3. Band. Nervensystem, Haut und Sinnesorgane.* Urban & Schwarzenberg, München – Wien – Baltimore, 1985: 365-394.

282. Petsche, H., Müller-Paschinger, I. B., Pockberger, H., Prohaska, O., Rappelsberger, P., and Vollmer, R.: Depth profiles of electrocortical activities and cortical architectonics. In: Brazier, M. A. B., and Petsche, H. (Eds.) *Architectonics of the Cerebral Cortex.* Raven Press, New York, 1978: 257-280.

283. Petsche, H., Pockberger, H., and Rappelsberger, P.: Current source density studies of epileptic phenomena and the morphology of the rabbit's striate cortex. In: Klee, M. R. (Ed.) *Physiology and Pharmacology of Epileptogenic Phenomena.* Raven Press, New York, 1982: 53-63.

284. Petsche, H., Pockberger, H., and Rappelsberger, P.: On the search of the sources of the EEG. *Neuroscience* 11: 1-27, 1984.

285. Petsche, H., Pockberger, H., and Rappelsberger, P.: Mechanisms leading to the propagation of self-sustained seizure activities. In: Wieser, H. G., Speckmann, E. J., and Engel, J. Jr. (Eds.) *Current Problems in Epilepsy: The Epileptic Focus.* John Libbey, London, 1987: 59-81.

286. Petsche, H., Prohaska, O., Rappelsberger, P., Vollmer, R., and Pockberger, H.: Simultaneous laminar intracortical recordings in seizures. *Electroenceph. Clin. Neurophysiol.* 42: 414-416, 1977.

287. Pfurtscheller, G., and Cooper, R.: Frequency dependence of the transmission of the EEG from the cortex to the scalp. *Electroenceph. Clin. Neurophysiol.* 38: 93-96, 1975.

288. Phillips, C. G., Powell, T. P. S., and Wiesendanger, M.: Projection from low treshold muscle afferents of hand and forearm to area 3a of baboon's cortex. *J. Physiol.* 217: 419-446, 1971.

289. Plonsey, R., and Heppner, D.: Considerations of quasistationarity in electro-physiological systems. *Bull. Math. Biophys.* 29: 657-664, 1967.

290. Pockberger, H., Petsche, H., and Rappelsberger, P.: Intracortical aspects of penicillin-induced seizure patterns in the rabbit's motor cortex. In: Speckmann,

E. J., and Elger, C. E. (Eds.) *Epilepsy and the Motor System.* Urban and Schwarzenberg, München, 1981: 161-178.

291. Pockberger, H., Rappelsberger, P., and Petsche, H.: Penicillin-induced epileptic phenomena in the rabbits neocortex I. The development of interictal spikes after epicortical application of penicillin. *Brain Res.* 309: 247-260, 1984.

292. Pockberger, H., Rappelsberger, P., and Petsche, H.: Penicillin-induced epileptic phenomena in the rabbits neocortex II. Laminar specific generation of interictal spikes after the application of penicillin to different cortical depths. *Brain Res.* 309: 261-269, 1984.

293. Powell, T. P. S., and Mountcastle, V. B.: The cytoarchitecture of the postcentral gyrus of the monkey Macaca mulatta. *Bull. Johns Hopkins Hosp.* 105: 108-131, 1959.

294. Powell, T. P. S., and Mountcastle, V. B.: Some aspects of the functional organization of the cortex of the postcentral gyrus of the monkey: a correlation of findings obtained in a single unit analysis with cytoarchitecture. *Bull. Johns Hopkins Hosp.* 105: 133-162, 1959.

295. Press, W. R., Flannery, B. P., Teukolsky, S. A., and Vetterling, W. T.: *Numerical Recipes.* Cambridge University Press, Cambridge, 1986.

296. Ragot, R. A., and Remond, A.: EEG field mapping. *Electroenceph. Clin. Neurophysiol.* 45: 417-421, 1978.

297. Rall, W.: Electrophysiology of a dendritic neuron model. *Biophysics J.* 2: 145-167, 1962.

298. Randolph, M., and Semmes, J.: Behavioral consequences of selective subtotal ablations in the postcentral gyrus of *Macaca Mulatta. Brain Res.* 70: 55-70, 1974.

299. Rappelsberger, P., Pockberger, H., and Petsche, H.: Current source density analysis: methods and application to simultaneously recorded field potentials of the rabbit's visual cortex. *Pflügers Arch. ges. Physiol.* 389: 159-170, 1981.

300. Rappelsberger, P., Pockberger, H., and Petsche, H.: The contribution of the cortical layers to the generation of the EEG: field potential and current source density analysis in the rabbit's visual cortex. *Electroenceph. Clin. Neurophysiol.* 53: 255-269, 1982.

301. Ricci, G. B., Romani, G. L., Salustri, C., Pizella, V., Buonomo, S., Peresson, M., and Modena, I.: Study of focal epilepsy by multichannel neuromagnetic measurements. *Electroenceph. Clin. Neurophysiol.* 66: 358-368, 1987.

302. Romani, G. L., and Leoni, R.: Localization of cerebral sources by neuromagnetic measurements. In: Weinberg, H., Stroink, G., and Katila, T. (Eds.) *Biomagnetism: Applications and Theory.* Pergamon Press, New York – Toronto, 1985: 205-220.

303. Romani, G. L., and Rossini, P.: Neuromagnetic functional localization: principles, state of the art, and perspectives. *Brain Topography* 1: 5-21, 1988.

304. Romani, G. L., Williamson, S. J., and Kaufman, L.: Biomagnetic instrumentation. *Rev. Sci. Instrum.* 53: 1815-1845, 1982.

305. Rose, D. F., Sato, S., Smith, P. D., Porter, R. J., Theodore, W. H., Friauf, W., Bonner, R., and Jabbari, B.: Localization of magnetic interictal discharges in temporal lobe epilepsy. *Ann. Neurol.* 22: 348-354, 1987.

306. Rose, D. F., Smith, P. D., and Sato, S.: Magnetoencephalography and epilepsy research. *Science* 238: 329-335, 1987.

307. Rosenthal, J., Walter, H. J., and Amassion, V. E.: An analysis of the activation of motor cortical necrosis by surface stimulation. *J. Neurophysiol.* 30: 844-858, 1967.

308. Rösler, R., and Manzey, D.: Principal components and varimax-rotated components in event-related potential research: Some remarks on their interpretation. *Biol. Psychol.* 13: 2-26, 1981.

309. Rossini, P. M., Cilli, M., Narici, L., Peresson, M., Pizzella, V., Romani, G. L., Salustri, C., Traversa, R., and DiLuzio, S.: Short latency somatosensory evoked activity to median nerve stimulation: differences in electric and magnetic scalp recordings. In: Atsumi, K., Kotani, M., Ueno, S., Katila, T., and Williamson, S. (Eds.) *Biomagnetism '87*. Tokyo Denki University Press, Tokyo, 1988:

310. Rossini, P. M., Gigli, G. L., Marciani, M. G., Zarola, F., and Caramia, M.: Non-invasive evaluation of input-output characteristics of sensorimotor cerebral areas in healthy humans. *Electroenceph. Clin. Neurophysiol.* 68: 88-100, 1987.

311. Rossini, P. M., Narici, L., Pizzella, V., Romani, G. L., and Traversa, R.: On the frontal components of somatosensory scalp responses to median nerve stimulation: neuromagnetic demonstration of an anterior, midline generator. In: Williamson, S. J., Hoke, M., Stroink, G., and Kotani, M. (Eds.) *Advances in Biomagnetism*. Plenum Press, New York – London, 1989: 157-160.

312. Ruch, T. C., Patton, H. D., Woodbury, J. W., and Towe, A. L.: *Neurophysiology, 2nd ed.* W.B. Saunders Company, Philadelphia, 1965.

313. Salar, G., Iob, I., and Mingrino, S.: Somatosensory evoked potentials before and after percutaneous thermocoagulation of the Gasserian ganglion for trigemianl neuralgia. In: Courjon, J., Mauguiere, F., and Revol, M. (Eds.) *Clinical Applications of Evoked Potentials in Neurology*. Raven Press, New York, 1982: 359-365.

314. Sarvas, J.: Basic mathematical and electromagnetic concepts of the biomagnetic inverse problem. *Phys. Med. Biol.* 32: 11-22, 1987.

315. Sato, S., Rose, D. F., and Kufta, C. V.: Localization of interictal spikes using a seven-channel magnetometer. In: Atsumi, K., Kotani, M., Ueno, S., Katila, T., and Williamson, S. (Eds.) *Biomagnetism '87*. Tokyo Denki University Press, Tokyo, 1988: 206-209.

316. Sato, S., Rose, D., and Porter, R.: Single magnetic spike mapping. In: Weinberg, H., Stroink, G., and Katila, T. (Eds.) *Biomagnetism: Applications and Theory*. Pergamon Press, New York, 1985: 261-263.

317. Sato, S., Sheridan, P., Smith, P., Bonner, R., Weinstock, H., Nissenoff, M., Rose, D., Theodore, W., Friauf, W., and Porter, R.: Comparison of EEG, MEG, and ECoG in epileptic patients. In: Weinberg, H., Stroink, G., and Katila, T. (Eds.) *Biomagnetism: Applications and Theory*. Pergamon Press, New York, 1985: 311-315.

318. Sato, S., and Smith, P. D.: Magnetoencephalography. *J. Clin. Neurophysiol.* 2: 173-192, 1985.

319. Scherg, M.: Spatio-temporal modelling of early auditory evoked potentials. *Rev. Laryng.* 105: 163-170, 1984.

320. Scherg, M.: Dipole source potentials of the auditory cortex in normal subjects and in patients with temporal lobe lesions. In: Grandori, F., Hoke, M., and Romani, G. L. (Eds.) *Auditory Evoked Magnetic Fields and Electric Potentials. Advances in Audiology. Vol. 6.* Karger, Basel, 1990: 165-193.

321. Scherg, M.: Fundamentals of dipole source potential analysis. In: Grandori, F., Hoke, M., and Romani, G. L. (Eds.) *Auditory Evoked Magnetic Fields and Electric Potentials. Advances in Audiology. Vol. 6.* Karger, Basel, 1990: 40-69.

322. Scherg, M., Hari, R., and Hämäläinen, M.: Frequency specific sources of the auditory N19-P30-P50 response detected by a multiple source analysis of evoked magnetic fields and potentials. In: Williamson, S. J., Hoke, M., Stroink, G., and Kotani, M. (Eds.) *Advances in Biomagnetism*. Plenum Press, New York, 1989: 97-100.

323. Scherg, M., Vajsar, J., and Picton, T. W.: A source analysis of the late human auditory evoked potentials. *J. Cog. Neurosci.* 1: 336-355, 1989.

324. Scherg, M., and Von Cramon, D.: A new interpretation of the generators of the BAEP waves I-V: results of a spatiotemporal dipole model. *Electroenceph. Clin. Neurophysiol.* 62: 290-299, 1985.

325. Scherg, M., and Von Cramon, D.: Two bilateral sources of the late AEP identified by a spatio-temporal dipole model. *Electroenceph. Clin. Neurophysiol.* 62: 32-44, 1985.

326. Scherg, M., and Von Cramon, D.: Evoked dipole source potentials of the human auditory cortex. *Electroenceph. Clin. Neurophysiol.* 65: 344-360, 1986.

327. Schlag, J.: Generation of brain evoked potentials. In: Thompson, R. F., and Patterson, M. M. (Eds.) *Bioelectric Recording Techniques. Cellular Processes and Brain Potentials, Part A.* Academic Press, New York, 1973: 273-316.

328. Schneider, S., Abraham-Fuchs, K., Daalmans, G., Folberth, W., Hoenig, H. E., Reichenberger, H., Röhrlein, G., Seifert, H., and Wirth, A.: Development and performance of a multichannel system for studies of biomagnetic signals of brain and heart. In: Williamson, S. J., Hoke, M., Stroink, G., and Kotani, M. (Eds.) *Advances in Biomagnetism.* Raven Press, New York – London, 1989: 669-672.

329. Slimp, J. C., Tamas, L. B., Stolov, W. C., and Wyler, A. R.: Somatosensory evoked potentials after removal of somatosensory cortex in man. *Electroenceph. Clin. Neurophysiol.* 65: 11-117, 1986.

330. Stefan, H., Schneider, S., Abraham-Fuchs, K., Bauer, J., Feistel, H., Pawlik, G., Neubauer, U., Röhrlein, G., and Huk, W. J.: Magnetic source localization in focal epilepsy. *Brain* 113: 1347-1359, 1990.

331. Steinmetz, H., Fürst, G., and Meyer, B.: Craniocerebral topography within the international 10-20 system. *Electroenceph. Clin. Neurophysiol.* 72: 499-506, 1989.

332. Stockard, J. J., and Iragui, V. J.: Clinically useful applications of evoked potentials in adult neurology. *J. Clin. Neurophysiol.* 1: 159-202, 1984.

333. Stöhr, M., and Petruch, F.: Somatosensory evoked potentials following stimulation of the trigeminal nerve in man. *J. Neurol.* 220: 95-98, 1979.

334. Stöhr, M., Petruch, F., and Scheglmann, K.: Somatosensory evoked potentials following trigeminal nerve stimulation in trigeminal neuralgia. *Ann. Neurol.* 9: 63-66, 1981.

335. Stohr, P. E., and Goldring, S.: Origin of somatosensory evoked scalp responses in man. *J. Neurosurg.* 31: 117-127, 1969.

336. Stohr, P. E., Goldring, S., and O'Leary, J. L.: Patterns of unit discharge associated with direct cortical response in monkey and cat. *Electroenceph. Clin. Neurophysiol.* 15: 661-669, 1963.

337. Stok, C. J.: *The Inverse Problem in EEG and MEG with Applications to Visual Evoked Responses.* PhD, University Twente, 1986.

338. Stok, C. J., Meijs, W. H., and Peters, M. J.: Inverse solutions based on MEG and EEG applied to volume conductor analysis. *Phys. Med. Biol.* 32: 99-104, 1987.

339. Strick, P. L., and Preston, J. B.: Multiple representations in the primate motor cortex. *Brain Res.* 154: 366-370, 1978.

340. Strick, P. L., and Preston, J. B.: Two representations of the hand in area 4 of a primate. I. Motor output organization. *J. Neurophysiol.* 48: 139-149, 1982.

341. Strick, P. L., and Preston, J. B.: Two representations of the hand in area 4 of a primate. II. Somatosensory input organization. *J. Neurophysiol.* 48: 150-159, 1982.

342. Suk, J., Cappell, J., Ribary, U., Yamamoto, T., and Llinas, R. R.: Magnetic localization of somatically evoked reponses in the human brain. In: Williamson, S. J., Hoke, M., Stroink, G., and Kotani, M. (Eds.) *Advances in Biomagnetism.* Plenum Press, New York – London, 1989: 165-168.

343. Sutherling, W. W.: *Protocol for Cortical Stimulations.* 1989.
344. Sutherling, W. W., and Barth, D. S.: Neocortical propagation in temporal lobe spike foci on magnetoencephalography and electroencephalography. *Ann. Neurol.* 25: 373-381, 1989.
345. Sutherling, W. W., Crandall, P., Cahan, L. D., and Barth, D. S.: The magnetic field of epileptic spikes agrees with intracranial localizations in complex partial epilepsy. *Neurology* 38: 778-786, 1988.
346. Sutherling, W. W., Crandall, P. H., Darcey, T. M., Becker, D. P., Levesque, M. F., and Barth, D. S.: The magnetic and electric fields agree with intracranial localizations of somatosensory cortex. *Neurology* 38: 1705-1714, 1988.
347. Sutherling, W. W., Crandall, P. H., Engel, J. Jr., Darcey, T. M., Cahan, L. D., and Barth, D. S.: The magnetic field of complex partial seizures agrees with intracranial localizations. *Ann. Neurol.* 21: 548-558, 1987.
348. Sutherling, W. W., Levesque, M. F., and Baumgartner, C.: Cortical sensory representation of the human hand: Size of finger regions and nonoverlapping digit somatotopy. *Neurology* 42: 1020-1028, 1992.
349. Sutherling, W. W., Risinger, M. W., Crandall, P. H., Baumgartner, C., Cahan, L. D., Barth, D. S., and Levesque, M. F.: Functional anatomy of the posterior dorsolateral frontal lobe seizure. *Neurology* 40: 87-98, 1990.
350. Swadlow, H. A., Rose, D. L., and Waxman, S. G.: Characteristics of interhemispheric impulse conduction between prelunate gyrus of the rhesus monkey. *Exp. Brain Res.* 33: 455-467, 1978.
351. Tanji, J., and Wise, S. P.: Submodality distribution in sensorimotor cortex of unanesthetized monkey. *J. Neurophysiol.* 45: 467-481, 1981.
352. Tiihonen, J., Hari, R., and Hämäläinen, M.: Early deflections of cerebral magnetic responses to median nerve stimulation. *Electroenceph. Clin. Neurophysiol.* 74: 290-296, 1989.
353. Tiihonen, J., Hari, R., Kajola, M., and Hämäläinen, M.: Evoked and spontaneous magnetic activity of the human somatosensory cortex. In: Williamson, S. J., Hoke, M., Stroink, G., and Kotani, M. (Eds.) *Advances in Biomagnetism.* Plenum Press, New York – London, 1989: 169-172.
354. Tomberg, C., Desmedt, J. E., Ozaki, I., Nguyen, T. H., and Chalkin, V.: Mapping somatosensory evoked potentials to finger stimulation at intervals of 450 to 4000 msec and the issue of habituation when assessing early cognitive components. *Electroenceph. Clin. Neurophysiol.* 74: 347-358, 1989.
355. Tomberg, C., Noel, P., Ozaki, I., and Desmedt, J.: Inadequacy of the average reference for the topographic mapping of focal enhancements of brain potentials. *Electroenceph. Clin. Neurophysiol.* 77: 259-265, 1990.
356. Towe, A. L.: On the nature of the primary evoked response. *Exp. Neurol.* 113-139, 1966.
357. Tracey, D. J., Asanuma, C., Jones, E. G., and Porter, R.: Thalamic relay to motor cortex: afferent pathways from brain stem, cerebellum, and spinal cord in monkeys. *J. Neurophysiol.* 44: 532-554, 1980.
358. Tsumoto, T., and Iwama, K.: Conduction velocities of leminiscal and thalamocortical fibers: their somatotopic differentiation and a rule of connection in the thalamic relay system. *Brain Res.* 44: 666-669, 1972.
359. Urasaki, E., Wada, S., Kadoya, C., Yokota, A., Matsuoka, S., and Shima, F.: Origin of scalp far-field N18 of SSEPs in response to median nerve stimulation. *Electroenceph. Clin. Neurophysiol.* 77: 39-51, 1990.
360. Van Rotterdam, A.: Limitations and difficultis in signal processsing by means of the principal component analysis. *IEEE Trans. Biomed. Engng.* BME-17: 266-269, 1970.

361. Vogt, M.: Über omnilaminäre Strukturdifferenzen und lineare Grenzen der architektonischen Felder der hinteren Zentralwindung des Menschen. *J. Psychol. Neurol.* 35: 177-193, 1928.

362. Vollmer, R., Petsche, H., Pockberger, H., Prohaska, O., and Rappelsberger, P.: Spatiotemporal analysis of seizure activities in a homogeneous cytoarchitectonic region. In: Brazier, M. A. B., and Petsche, H. (Eds.) *Architectonics of the Cerebral Cortex.* Raven Press, New York, 1978: 281-305.

363. Wastell, D. G.: PCA and varimax rotation: Some comments on Rösler and Manzey. *Biol. Psychol.* 13: 27-29, 1981.

364. Waxman, S. G., and Bennet, M. V.: Relative conduction velocities of small myelinated and non-myelinated fibers in the central nervous system. *Nature* 338: 217-219, 1972.

365. Werner, G., and Whistel, B. L.: Functional organization of the somatosensory cortex. In: Iggo, A. (Ed.) *Handbook of Sensory Physiology, Somatosensory System.* Springer, Berlin, 1973: 621-700.

366. Wieser, H. G.: Selective amygdalohippocampectomy: indications, investigative technique and results. In: Symon, L., Brihaye, J., Guidetti, B., Loew, F., Miller, J. D., Nornes, H., Pasztor, E., Pertuiset, B., and Yasargil, M. G. (Eds.) *Advances and Technical Standards in Neurosurgery, Vol. 13.* Springer, Vienna, 1986: 39-133.

367. Wieser, H. G., and Yasargil, M. G.: Selective amygdalohippocampectomy as a surgical treatement of mesiobasal limbic epilepsy. *Surg. Neurol.* 17: 445-457, 1982.

368. Wikswo, J. P., and Roth, B. J.: Magnetic determination of the spatial extent of a single cortical current source: a theoretical analysis. *Electroenceph. Clin. Neurophysiol.* 69: 266-276, 1988.

369. Williamson, S. J., and Kaufman, L.: Biomagnetism. *J. Magn. Magn. Mat.* 22: 129-201, 1981.

370. Williamson, S. J., and Kaufman, L.: Magnetic fields of the cerebral cortex. In: Erné, S. N., Hahlbohm, H. H., and Lübbig, H. (Eds.) *Biomagnetism.* Walter de Gruyter, Berlin, 1981: 353-402.

371. Williamson, S. J., Robinson, S. E., and Kaufman, L.: Methods and instrumentation for biomagnetism. In: Atsumi, K., Kotani, M., Ueno, S., Katila, T., and Williamson, S. (Eds.) *Biomagnetism '87.* Tokyo Denki University Press, Tokyo, 1988: 18-25.

372. Wood, C. C.: Application of dipole localization methods to source identification of human evoked potentials. *Ann. NY Acad. Sci.* 388: 139-155, 1982.

373. Wood, C. C., and Allison, T.: Interpretation of evoked potentials: a neurophysiological perspective. *Can. J. Psychol.* 35: 113-135, 1981.

374. Wood, C. C., Allison, T., Goff, W. R., Williamson, P. D., and Spencer, D. D.: Localization of human sensorimotor cortex during surgery by cortical surface recording of somatosensory evoked potentials. *Electroenceph. Clin. Neurophysiol.* 51: 36P-37P, 1981.

375. Wood, C. C., Cohen, D., Cuffin, B. N., Yarita, M., and Allison, T.: Electrical sources in human somatosensory cortex: identification by combined magnetic and potential recordings. *Science* 227: 1051-1053, 1985.

376. Wood, C. C., and McCarthy, G.: Principal component analysis of event related potentials: simulation studies demonstrate misallocation of variance across components. *Electroenceph. Clin. Neurophysiol.* 59: 249-260, 1984.

377. Wood, C. C., McCarthy, G., and Darcey, T. M.: Principal component analysis and dipole localization models applied to surface potentials generated by multiple dipoles. *Electroenceph. Clin. Neurophysiol.* 64: 80P, 1986.

378. Wood, C. C., Spencer, D. D., Allison, T., McCarthy, G., Williamson, P. D., and

Goff, W. R.: Localization of human sensorimotor cortex during surgery by cortical surface recording of somatosensory evoked potentials. *J. Neurosurg.* 68: 99-111, 1988.

379. Wood, C. C., and Wolpaw, J. R.: Scalp distribution of human auditory evoked potentials. II. Evidence for overlapping sources and involvement of auditory cortex. *Electroenceph. Clin. Neurophysiol.* 54: 25-38, 1982.

380. Woolsey, C. N.: Organization of somatic sensory and motor areas of the cerebral cortex. In: Harlow, H. F., and Woolsey, C. N. (Eds.) *Biological and Biochemical Bases of Behaviour.* University of Wisconsin Press, Madison, 1958: 63-81.

381. Woolsey, C. N.: *Cortical Sensory Organization.* Humana Press, Clifton, 1981.

382. Woolsey, C. N., and Erickson, T. C.: Study of postcentral gyrus of man by the evoked potential technique. *Trans. Am. Neurol. Assoc.* 75: 50-52, 1950.

383. Woolsey, C. N., Erickson, T. C., and Gilson, W. E.: Localization in somatic sensory and motor areas of human cortex determined by direct recording of evoked potentials and electrical stimulation. *J. Neurosurg.* 51: 476-506, 1979.

384. Woolsey, C. N., Walker, A. N., and Erickson, T. C.: Somatic afferent representation in the cerebral cortex of man. *Proceedings of the Fourth International Congress of Neurology* 70-71, 1949.

385. Yamada, T., Kayamori, R., Kimura, J., and Beck, D. O.: Topography of somatosensory evoked potentials after stimulation of the median nerve. *Electroenceph. Clin. Neurophysiol.* 59: 29-43, 1984.

386. Yamada, T., Kimura, J., and Nitz, D. M.: Short-latency somatosensory evoked potentials following median nerve stimulation in man. *Electroenceph. Clin. Neurophysiol.* 48: 367-376, 1980.

387. Yamada, T., Kimura, J., Wilkinson, J. T., and Kayamori, R.: Short- and long-latency median somatosensory evoked potentials. *Arch. Neurol.* 40: 215-220, 1983.

388. Zeitlhofer, J., Mamoli, B., Baumgartner, C., and Mayr, N.: Somatosensorisch evozierte Potentiale nach Tibialisstimulation. *Wien. Klin. Wschr.* 100: 6-11, 1988.

389. Zeitlhofer, J., Steiner, M., Bousek, K., Fitzal, S., Asenbaum, S., Wolner, E., and Deecke, L.: The influence of temperature on somatosensory-evoked potentials during cardiopulmonary bypass. *Eur. Neurol.* 30: 284-290, 1990.

390. Zeitlhofer, J., Steiner, M., Zadrobilek, E., Häusl, E., Sporn, P., Asenbaum, S., Oder, W., Baumgartner, C., and Deecke, L.: Evozierte Potentiale zur Verlaufs- und Prognosebeurteilung von Schädel-Hirn-Trauma-Patienten. *Anaesthesist* 38: 10-15, 1989.

391. Zimmerman, I. D.: A triple representation of the body surface in the sensorimotor cortex of the squirrel monkey. *Exp. Neurol.* 20: 415-431, 1968.

392. Zimmerman, J. E.: Evaluation of the SQUID and its use in biomagnetic research. In: Williamson, S. J., Hoke, M., Stroink, G., and Kotani, M. (Eds.) *Advances in Biomagnetism.* Plenum Press, New York – London, 1989: 67-72.